Contents

1. Introduction 3
   1.1 Globular clusters 3
   1.2 Red giants 4
   1.3 Extra-mixing in stars: observations 6
   1.4 Extra-mixing in stars: theoretical models 9
   1.5 A combined scenario 14
   1.6 Sites of primordial nucleosynthesis in globular clusters 16
   1.7 Possible role of the cluster medium 18
   1.8 Structure of the review 19
2. Extra-mixing in red giants 22
   2.1 Diffusion model for extra-mixing in globular-cluster red giants (1994–1995) 22
      2.1.1 Introduction 22
      2.1.2 Models and computation method 28
      2.1.3 Results 36
      2.1.4 Main conclusions 50
   2.2 A possible mechanism for extra-mixing in globular-cluster red giants (1999) 50
      2.2.1 Introduction 50
      2.2.2 Semi-empirical diffusion model for extra-mixing 55
      2.2.3 The Zahn extra-mixing mechanism 59
      2.2.4 Choice of criterion for shear instability 65
      2.2.5 Results of numerical computations 71
      2.2.6 Conclusion 79
   2.3 Episodic production of lithium in red giants (2000) 81
      2.3.1 Introduction 81
      2.3.2 Extra-mixing in red giants 83
      2.3.3 The problem of lithium-rich red giants 87
      2.3.4 Proposed solution to the lithium problem 90
      2.3.5 Conclusion 94
3. Chemical evolution of globular clusters 95
   3.1 A combined scenario: inherited chemical anomalies plus extra-mixing in red giants (1997) 95
      3.1.1 Introduction 95

3.1.2	Computer code used	99
3.1.3	Deep-mixing scenarios	102
3.1.4	Primordial nucleosynthesis scenario: inherited anomalies	118
3.1.5	Conclusion	131
3.2	Contribution of ^{26}Al to the anticorrelation between the O and Al abundances in globular-cluster red giants(2001)	132
3.2.1	Introduction	132
3.2.2	The problem of the O deficit	134
3.2.3	The problem of the Al excess	138
3.2.4	Concluding remarks	139
4.	Transition of MS stars with masses of 10 and 30 $M_\odot$ to states with stationary rotation (1998)	143
4.1	Introduction	143
4.2	Basic equations	146
4.3	Additional assumptions and simplifications	149
4.4	Calculation results	152
4.5	Main conclusions	157
5.	Conclusion	161
References		165
Index		173

ASTROPHYSICS AND SPACE PHYSICS REVIEWS

Editor: R.A. Sunyaev
Institute for Space Research
Russian Academy of Sciences
Moscow, Russia

Advisory Editor: M. Longair
Cavendish Laboratory
University of Cambridge
UK

GENERAL INFORMATION

Aims and Scope
Astrophysics and Space Physics Reviews publishes review articles covering significant developments in astronomy, theoretical astrophysics, cosmology, high energy astrophysics and space research in the former Soviet Union. Plans for future space experiments are also highlighted.

©2004 Cambridge Scientific Publishers. All rights reserved.

Except as permitted under national laws or under the photocopy license described below, no part of this publication may be reproduced or transmitted in any form or by any means, electronic, mechanical, photocopying or otherwise, or stored in a retrieval system of any nature, without the advance written permission of the Publisher.

Ordering Information
Each volume consists of an irregular number of parts depending upon size. Issues are available individually as well as by subscription. 2003 Volume: 12.

Orders may be placed with your usual supplier or at one of the addresses shown below. Journal subscriptions are sold on a per volume basis only. Claims for nonreceipt of issues will be honored if made within three months of publication of the issue. Subscriptions are available for microform editions; details will be furnished upon request. All issues are dispatched by airmail throughout the world.

Subscription Rates
Base list subscription price per volume: EUR 182.00. * This price is available only to individuals whose library subscribes to the journal OR who warrant that the journal is for their own use and provide a home address for mailing. Orders must be sent directly to the Publisher and payment must be made by personal check or credit card. Separate rates apply to academic and corporate/government institutions. Postage and handling charges are extra.

*EUR (Euro). The Euro is the worldwide base list currency rate; payment can be made by draft drawn on EURO currency in the amount shown, or in any other currency within the eurozone at the eurodenominated rate. All other currency payments should be made using the current conversion rate set by the Publisher. Subscribers should contact their agents or the Publisher. All prices are subject to change without notice.

Orders should be placed through the Publisher at the following addresses:

Cambridge Scientific Publishers
P.O. Box 806
Cottenham
Cambridge
CB4 8RT
UK
Tel: +44 (0)1954 251283
Fax: +44 (0)1954 252517
Email: janie.wardle@cambridgescientificpublishers.com
Website: www.cambridgescientificpublishers.com

Rights and Permissions/Reprints of Individual Articles
Permission to reproduce and/or translate material contained in this journal must be obtained in writing from the Publisher. This publication and each of the articles contained herein are protected by copyright. Except as allowed under national "fair use" laws, copying is not permitted by any means or for any purpose, such as for distribution to any third party (whether by sale, loan, gift or otherwise); as agent (express or implied) of any third party; for purposes of advertising or promotion; or to create collective or derivative works. A photocopy license is available from the Publisher for institutional subscribers that need to make multiple copies of single articles for internal study or research purposes. any unauthorized reproduction, transmission or storage may result in civil or criminal liability.

Copies of individual articles may be ordered through the Publisher's document delivery service. Please contact one of the addresses listed above.

Printed in UK

April 2004

DEEP MIXING IN GLOBULAR-CLUSTER RED GIANTS

P.A. Denissenkov

St. Petersburg State University, St. Petersburg

ABSTRACT

This review presents the results of studies of the origin of star-to-star abundance variations in globular clusters. In particular, peculiar abundances in red giants are ascribed to deep mixing penetrating the hydrogen burning shell. Two models of deep mixing are considered: a parametric diffusion model and a mechanism of rotationally induced meridional circulation and turbulent diffusion. To explain all the abundance anomalies observed in globular clusters the author proposes a scenario combining primordial nucleosynthesis and deep mixing in red giants. This combined scenario is used as the basis for a model of the chemical evolution of globular clusters. The review provides a useful reference text for graduates and researchers in astrophysics.

1. Introduction

1.1 Globular Clusters

Globular clusters are among the oldest known stellar aggregates in the Universe. The most recent age estimates for the oldest globular clusters yield $(11\pm 1.5)\cdot 10^9$ yrs [162], in good agreement with the time since the Big Bang, based on the recently refined value for the Hubble constant [155]. In our Galaxy, globular clusters are members of Population II and of the halo population. One of their distinguishing features is low metallicity ($[{\rm Fe/H}]$[1] < 0) compared to the solar value. The masses of globular clusters (10^5–$10^6\ M_\odot$)[2] are strikingly close to the Jeans mass, corresponding to the onset of the development of gravitational instability in an early stage of formation of the Galaxy. This leads to the natural conclusion that globular clusters are most likely to have separated from the protogalaxy before the formation of the Galactic disc [74]. The gravitational contraction of clouds and proto-clusters occurred on a timescale which is much shorter than the current ages of globular clusters. Consequently, all stars in an individual globular cluster can with high accuracy be considered to have the same age. Observations show that virtually all globular clusters (excluding only M22 and ω Cen) have the homogeneous chemical composition if we adopt iron as a test element. For example, Geisler and Sarajedini [82] found a very small scatter (0.03–0.09) in [Fe/H] (within a single cluster) in a sample of 12 globular clusters.

The evolutionary status of a star is completely determined by its initial mass M, chemical composition (the relative mass contents of hydrogen X and of "metals" – elements heavier than helium, Z) and age, of which only the first parameter varies from star to star within a globular cluster. Thus, it is the initial mass of a star that determines on which evolutionary branch in the "colour–magnitude" diagram it is located – on the main sequence (MS), sub-giant branch, red giant branch, horizontal branch, or asymptotic giant branch (AGB). In the

---

[1] We use the standard spectroscopic notation $[{\rm A/B}] = \lg[N({\rm A})/N({\rm B})]_{\rm star} - \lg[N({\rm A})/N({\rm B})]_\odot$, where $N({\rm A})$ and $N({\rm B})$ are the number densities of nuclides A and B, respectively.

[2] The main parameters for a large number of globular clusters can be found at the internet site http://physun.physics.mcmaster.ca/Globular.html.

majority of globular clusters, stars with $M \geq$1–2 $M_\odot$ have already finished their lives, so the colour–magnitude diagrams of globular clusters can only be used to investigate the evolution of low-mass stars.

Fortunately, there is another stellar "parameter" that contains information not only about the evolution of stars with the entire range of masses found in globular clusters, but also about the previous chemical evolution and origin of any individual cluster. This "parameter" is the *detailed* chemical composition of the cluster stars; i.e. as complete as possible a set of surface elemental abundances for individual stars, and especially for red giants (which are some orders of magnitude more luminous than MS stars, making them much easier to study using spectral analyses). The main goal of this review is to analyse and interpret this information. In order to achieve this goal, we must solve a whole set of related problems. These include: *(i)* developing physical models and corresponding computational algorithms describing extra-mixing in red giants; *(ii)* carrying out detailed calculations of nucleosynthesis in AGB stars; and *(iii)* constructing models for the chemical evolution of globular clusters, taking into account the products of nucleosynthesis created by AGB stars and Type II supernovae. As will be shown, all these nucleosynthesis and mixing processes have left their "imprints" in the current chemical compositions of globular cluster stars.

1.2 Red Giants

The most prolonged phase in the evolution of a star is its life on the main sequence, when its luminosity is provided by thermonuclear reactions that transform hydrogen into helium. The specific sequence of reactions and resulting stellar structure are determined by the temperature at the centre, which, in turn, depends on the mass of the star.

For example, an MS star of intermediate mass ($2.5 \leq M/M_\odot \leq 8$) has a convective core, in which CNO-cycle reactions occur surrounded by a radiative envelope. (For a detailed description of hydrostatic hydrogen burning cycles, see [4].) According to standard stellar evolution theory, the products of these reactions (He and changed proportions of CNO nuclides) make their way to the stellar surface and

DEEP MIXING IN GLOBULAR-CLUSTER RED GIANTS   5

become accessible to spectral observations only in the red giant (or supergiant) phase, after the star has left the main sequence. In a red giant, mixing is brought about by convection in the extended envelope, whose base reaches layers in which nuclear reactions occurred on the main sequence. This phenomenon marks the well-known "first dredge-up" in standard theory. The main source of energy of a red giant are CNO-cycle reactions occurring in a thin layer of hydrogen at the surface of the helium core (the so-called "hydrogen burning shell"). In the course of the subsequent evolution of a star with intermediate mass, there is a quiet burning of He at its centre, which begins to be transformed into C and O in the convective core. After He is depleted in the central regions of the star, the C–O core begins to be compressed, the electron gas in it degenerates, and the star moves to the asymptotic giant branch.

The AGB star consists of a degenerate carbon–oxygen core surrounded by (in order): a helium burning shell, a helium-rich intermediate layer, and a hydrogen burning shell. Around all these layers is a deep convective envelope. During the star's evolution on the AGB, thermal instabilities leading to flashes arise in the helium burning shell roughly every 10^4 years. During such flashes, the burning in the hydrogen shell dies down as a consequence of the expansion and cooling of lower-lying layers of stellar material (between flashes, this shell serves as the main supplier of energy). Another nucleosynthesis process encountered in AGB stars with masses $M \approx 5$–$7\,M_\odot$ is nuclear burning at the base of the convective envelope ("hot bottom burning") [113]. This can lead to the transformation of C into N, and also to Li synthesis [160], in full agreement with observations [179].

In low-mass MS stars ($M \leq 2.5\,M_\odot$, for example, the Sun), the transformation of H to He occurs in pp-chain reactions. The energy-releasing core is in radiative equilibrium. After the depletion of H in the centre, the low-mass star, like intermediate-mass stars, leaves the main sequence and becomes a red giant. One difference is that the helium core of the low-mass red giant is degenerate. As in intermediate-mass giants, the CNO cycle transforms H into He in a narrow layer at the surface of the helium core. This hydrogen burning shell is separated from the convective envelope by a radiative zone. In spite of its very small mass (1–5% of the total mass of the star), the radiative zone has a linear size up to 1–$2\,R_\odot$. In the course of the subsequent evolution of the low-mass star, an explosive onset of He burning occurs in its centre, called the "core helium flash". This is a consequence of the strongly degenerate electron gas there. As a result, the star moves to the horizontal branch. After He burning in the convective core, a degenerate CO core forms, and the low-mass star leaves the horizontal branch and moves to the AGB.

During the course of the star's evolution, the location of the hydrogen burning shell moves outward, toward the discontinuity in chemical composition that has formed at the base of the convective envelope at the end of the first dredge-up. An important point for our subsequent discussion is that only in low-mass red giants does the hydrogen burning shell cross this jump in chemical composition before helium burning is ignited in the star (i.e. while the star is still on the red giant branch). This is important since it is believed that the gradient in the mean molecular weight hinders extra-mixing (see, for example, Section 2.2 of this review). After the hydrogen burning shell has crossed the jump in chemical composition, this gradient remains close to zero in almost the whole radiative zone, from the base of the convective envelope to regions adjacent to the shell source. It is interesting in this connection that peculiarities in surface chemical composition that have been attributed to the action of extra-mixing are observed only in low-mass red giants [42, 45].

These evolutionary variations pertain only to single stars. In close binary systems, an appreciable influence on the evolution of a star is exerted by interaction with its companion. Late phases in the evolution of high-mass stars ($M \geq 30\,M_\odot$) are also more complex and ambiguous.

1.3 Extra-Mixing in Stars: Observations

Following well-established tradition, we will take "extra-mixing" in stars to refer to any mixing process distinct from ordinary convection, semi-convection, and convective overshooting. Sometimes, extra-mixing is still called "non-standard" mixing, since it is usually not included in standard calculations of stellar evolution. This omission in the standard theory can lead to serious errors when the results are used to model galactic chemical evolution, analyse stellar populations in nearby galaxies, construct isochrones used to determine the ages of stellar clusters, and in other astrophysical applications. The presence of extra-mixing has been most convincingly demonstrated in intermediate- and high-mass MS stars (bright O and B stars), and also in low-metallicity red giants, in globular-cluster red giants, in particular. In all these objects, CNO nuclides play the role of catalysts in the transformation of H to He in the CNO cycle.

In the last two decades, a great deal of observational data suggesting the action of some extra-mixing in the radiative zones of OB stars has been accumulated (see, for example, the review by Lyubimkov [126]). The most convincing evidence for this is the presence of N and He excesses in the atmospheres of OB stars, which are correlated with the mass and age of the star [123, 124, 83].

It has been shown that anomalies in He abundances in bright OB stars are accompanied by mass discrepancies, in the sense that the "evolutionary masses" (determined by comparing the position of the star on the Hertzsprung–Russell (HR) diagram with theoretical evolutionary tracks) are systematically higher than the "spectroscopic masses" [91, 90]. Langer [113], Weiss [211] and Denissenkov [58] have proposed that both the helium anomalies and these mass discrepancies could have the same origin – extra-mixing (of an as yet unknown nature) in the radiative zones of OB stars. Indeed, stellar material enriched in helium becomes more transparent to radiation, which could lead to an increase in stellar luminosity, resulting in overestimation of the evolutionary masses. However, in our view, a more important observational detail to which we should turn our attention is the delay in the enrichment of OB-star atmospheres in helium. Lyubimkov [126] found that the He abundances in OB stars begin to grow only after about 50% of the star's lifetime on the main sequence.

Stars that have left the main sequence in modern globular clusters have masses $M \approx 0.8 - 0.9\,M_\odot$ and metallicities $-2.4 \leq [\text{Fe/H}] \leq -0.2$, which corresponds to heavy-element abundances $8 \cdot 10^{-5} \leq Z \leq 0.01$ (we assume $[\text{Fe/H}] = \lg(Z/Z_\odot)$ and $Z_\odot = 0.01886$ [2]). Standard evolutionary calculations for Population II stars with $M = 0.8\,M_\odot$ predict very small variations in the surface abundances of ^{12}C, ^{13}C and ^{14}N in the first dredge-up: depending on the value of Z, we have $(Z, {}^{12}\text{C}/{}^{13}\text{C}, \Delta\lg{}^{12}\text{C}, \Delta\lg{}^{14}\text{N}) = (10^{-4}, 64, -0.0024, 0.0025)$, $(5 \cdot 10^{-4}, 50, -0.0064, 0.012)$, $(5 \cdot 10^{-3}, 45, -0.0092, 0.021)$ [66]. These values were obtained for initial solar abundances varied in proportion to $Z/Z_\odot$ (in particular, initially, $^{12}C/^{13}C = 90$). Here and below, the chemical symbol of a nuclide is also used to denote its number density.

When comparing these theoretical predictions with observational data on the chemical compositions of globular-cluster red giants, to our surprise, we discover many discrepancies: *(i)* as a rule, the observed ratio $^{12}C/^{13}C$ is very low, often approaching the equilibrium value of 3.5; *(ii)* differences in the C and N abundances among red giants in the same globular cluster reach an order of magnitude and more; *(iii)* the most intriguing result is that large (to 1 dex) differences in the O and Na abundances are observed among the red giants in many globular clusters, and the Al and Mg abundances also vary from star to star in individual clusters.

It is interesting that all the nuclides indicated above (^{12}C, ^{13}C, N, O, Na, Mg and Al) participate in hydrostatic hydrogen burning [5, 4]. Like CNO nuclides in the CNO cycle, Ne, Na, Mg and Al play the role of catalysts in the NeNa and MgAl cycles. Although the relative abundances of catalysts can vary in nuclear reactions, their total abundance should remain constant. In spite of the large variation in the individual abundances of C, N and O, an approximate constancy in the total C+N+O abundances is, indeed, observed in the globular clusters M13 and M3 [178], in NGC 362 and NGC 288 [70] and in ω Cen [139]. There is also evidence that the total Mg+Al abundance in M13 is constant [107].

In contrast to the situation for oxygen, the abundances of heavier α-process elements such as Si, Ca and Ti, which are produced during the consecutive capture of α particles in massive stars and do not participate in hydrostatic hydrogen burning, do not vary from star to star in globular clusters. Instead, the mean abundances of α-process elements in globular clusters are in fairly close agreement with the corresponding abundances in Population II field dwarfs, $\langle[\alpha/Fe]\rangle \approx +0.4$ [215].

Direct confirmation of the presence of extra-mixing in red giants could be provided by the detection of evolutionary surface abundance variations. Since the evolution of a star on the red giant branch is accompanied by a monotonic (and quite appreciable) increase in its luminosity in the presence of almost constant effective temperature, it could be useful to search for correlations between abundance and luminosity (or absolute magnitude). Such correlations have, in fact, been found. For example, in 1986, Langer *et al.* [117] discovered that, in red giants in the globular cluster M92, the quantity [C/Fe] decreases with decreasing M_V (i.e. with increasing luminosity). This result was recently verified and confirmed in [10]. Further, Gratton *et al.* [87] have shown that, in field red giants with well-known luminosities and metal abundances in the interval $-2 \leq$ [Fe/H] ≤ -1, evolutionary variations in the surface abundances of Li, C, N and in the ratio ^{12}C/^{13}C are observed after the end of the first dredge-up. Thus, the presence of extra-mixing in low-metallicity red giants is currently beyond doubt. However, the questions of the depth of this mixing and its physical nature remain open.

1.4 Extra-Mixing in Stars: Theoretical Models

Let us summarise the above discussion: the observational data confirm that star-to-star variations in the chemical abundances of globular-cluster red giants arise in the presence of hydrostatic hydrogen burning. We must now try to answer the main question – where does this hydrogen burning occur? We will call theoretical models proposing that this burning occurs in the hydrogen burning shell in these red giants themselves, with extra-mixing connecting the burning shell with the base of the convective envelope, the "evolutionary scenario" or "mixing scenario". An alternative model proposing that the anomalous abundances observed in globular-cluster red giants were inherited from the cluster interstellar medium during or after their formation we will call the "primordial nucleosynthesis scenario". In this latter model, it is thought that the required hydrogen burning occurred in previous generations of stars, for which the most likely candidates are intermediate-mass AGB stars [66].

In the late 1970s, abundance variations from star to star in globular clusters were known only for C, N and O (see the brief review by Da Costa [52]). To explain these observational data, Sweigart and Mengel [192] proposed an evolutionary scenario in which the extra-mixing was identified with classical meridional circulation [71, 206, 190]. After the first reports by Cottrell and Da Costa [49] and Norris *et al.* [137] that giants in NGC 6752 with N overabundances also have enhanced Na and Al abundances, data began to be accumulated in the 1980s suggesting that this strange correlation of N with

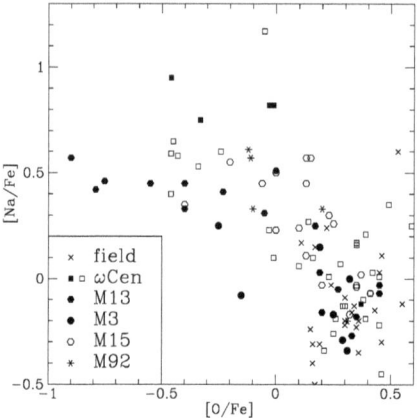

Figure 1.1. Global anticorrelation of [O/Fe] and [Na/Fe] (in the terminology of Kraft [103]).

Na and Al was a general property of many globular clusters. Moreover, Paltoglou and Norris [142] found that the Na abundance was anticorrelated with the O abundance in ω Cen, anticipating the later discovery of the "global anticorrelation" between [O/Fe] and [Na/Fe] by Kraft et al. [106] (Fig. 1.1).

Before 1990, data on overabundances of Na and Al in globular-cluster red giants were considered weighty evidence for primordial nucleosynthesis scenarios, since it was not understood at that time how these fairly heavy elements could form during hydrogen burning in stars at early stages of their evolution.

Mixing scenarios were "rehabilitated" by low-energy resonances in reactions of the NeNa and MgAl cycles. In 1990, Denissenkov and Denissenkova [61] first showed that the temperature in the outer part of the hydrogen burning shell in a red giant was sufficiently high for the reaction ^{22}Ne(p,γ)^{23}Na to occur there (thanks to resonance!) even more rapidly than the burning of O in the CNO cycle. Further, Langer et al. [115] established that ^{27}Al can also be synthesized (primarily from ^{25}Mg) slightly below the layer in which the O abundance

DEEP MIXING IN GLOBULAR-CLUSTER RED GIANTS 11

begins to fall off, especially in giants with very low metallicities (see also [64]).

We emphasise that Denissenkov and Denissenkova [61] essentially predicted a step-like form for the global anticorrelation of [O/Fe] with [Na/Fe]. Their pioneering work stimulated a broad range of observational programmes concerned with detailed analyses of the chemical composition of globular-cluster red giants in the United States (at the Lick and McDonald Observatories), Australia (Mount Stromlo and Siding Spring Observatories) and (quite recently) Europe (European Southern Observatory). These programmes are ongoing.

Extra-mixing in red giants is usually modelled by diffusion [43, 64]. More refined models are also sometimes used, such as models with two-flow "conveyor-belt" circulation [161]. A number of studies have been restricted to simple discussions of abundance profiles near the hydrogen source, or draw some conclusions based on analyses of abundance variations with time in the presence of constant temperature and density [38, 116, 191]. As a rule, the physical mechanism for the extra-mixing is not discussed; instead, several free parameters whose values are chosen to fit the observational data are introduced. For example, the diffusion model of Denissenkov and Weiss [64] has two parameters: the relative depth δM_{mix} (equal to zero at the boundary between the helium core and the hydrogen burning shell and to unity at the base of the convective envelope) and the extra-mixing rate (diffusion coefficient) D_{mix} (Fig. 1.2). This model is able to reproduce the global anticorrelation between [O/Fe] and [Na/Fe] fairly well with the values $\delta M_{\mathrm{mix}} = 0.06 - 0.07$, $D_{\mathrm{mix}} = 1 - 5 \cdot 10^9$ cm$^2 \cdot$s^{-1} [64]. It is able to simultaneously explain four of the correlations between the C, N, O, Na and Al abundances for giants in clusters ω Cen and M4 using a single set of parameters (for each cluster) to describe the extra-mixing (see [66] and Sections 3.1 and 3.2 of this review). Note that the extent of the global anticorrelation for very low O abundances ([O/Fe] < -0.4) is determined entirely by data for the cluster M13, which is unique in this sense (see Fig. 1.1). Analysis of the results obtained using this diffusion model led Denissenkov and Tout [63] to the following important conclusions. The extent of the global anticorrelation in the vertical direction (i.e. the range of [Na/Fe] values) depends primarily on the mixing depth. On the other hand, the extent in the horizontal direction (the range of [O/Fe])

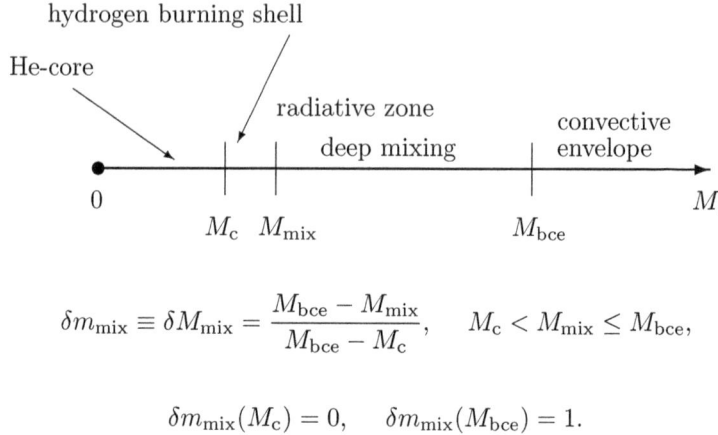

$$\delta m_{\text{mix}} \equiv \delta M_{\text{mix}} = \frac{M_{\text{bce}} - M_{\text{mix}}}{M_{\text{bce}} - M_{\text{c}}}, \quad M_{\text{c}} < M_{\text{mix}} \leq M_{\text{bce}},$$

$$\delta m_{\text{mix}}(M_{\text{c}}) = 0, \quad \delta m_{\text{mix}}(M_{\text{bce}}) = 1.$$

Figure 1.2. Structure of a red giant with deep mixing in the radiative zone.

provides information first and foremost about the mixing rate. In this sense, globular-cluster red giants can be considered a unique laboratory and the global anticorrelation itself an effective tool for testing various mechanisms for extra-mixing in stars. In particular, the morphology of the global anticorrelation suggests that the depth of the extra-mixing is probably approximately the same in all globular clusters (with the possible exception of ω Cen), while the mixing rates differ, with giants in M13 undergoing the most rapid mixing.

Denissenkov and Tout [63] recently proposed the first realistic physical mechanism for extra-mixing in red giants – that of Zahn [219, 132]. In this mechanism, extra-mixing in the radiative zone of a star is taken to result from the joint action of meridional circulation and turbulent diffusion arising due to rotation. A brief description of this mechanism is given below.

In a rotating star, the surfaces of constant entropy do not coincide with equipotential surfaces. Consequently, a spherically symmetrical flow of radiative energy is not able to support a state of thermal equilibrium, giving rise to meridional circulation. The resulting

DEEP MIXING IN GLOBULAR-CLUSTER RED GIANTS 13

redistribution of angular momentum causes the initially uniform rotation to become differential. It is supposed that the turbulence arising due to hydrodynamic instabilities associated with the differential rotation is very anisotropic, with the horizontal component of the turbulent viscosity appreciably exceeding the vertical component. This main suggestion by Zahn brings about a rapid transition to a state of shellular rotation, in which the angular velocity depends only on distance from the centre of the star. Thus, the entire problem becomes one-dimensional. Among various hydrodynamical instabilities, the shear (Kelvin–Helmholtz) instability has one of the shortest growth times. The radial thermal structure of the radiative zone suppresses the growth of perturbations, and corresponding instability criteria must be used in order to study their possible development. If perturbations are able to grow, the resulting (vertical) turbulent diffusion begins to compete with meridional circulation in the redistribution of angular momentum. On the other hand, there is nothing to prevent the growth of perturbations on equipotential surfaces, leading to the development of powerful horizontal turbulent diffusion, which smooths inhomogeneities in the rotation on these surfaces. This justifies Zahn's main assumption that the turbulence is very anisotropic.

In [63], Denissenkov and Tout developed an efficient algorithm and corresponding computer code for the solution of the non-linear partial differential equation (with radius derivatives up to the fourth order) describing the redistribution of angular momentum by meridional circulation and turbulent diffusion. The structure of this equation is such that, under stationary conditions, the meridional circulation and turbulent diffusion cancel each other out after a short time (the thermal time). This occurs, for example, in the radiative envelopes of OB stars. As a result, the rotation-induced extra-mixing in OB stars proves to be very slow, in spite of their rapid rotation. This qualitatively explains the delay in the appearance of He overabundances in the atmospheres of OB stars [67].

Denissenkov and Tout [63] showed that, in the radiative zone of a red giant, Zahn extra-mixing between the hydrogen burning shell and the base of the convective envelope does not appreciably affect the quasi-static inward flow of hydrogen-rich material in this region (which feeds the shell source). One consequence is a fairly steep angular velocity profile ($\Omega \propto r^{-2}$). Thus, the radiative zone should be in a state of strong differential rotation, so that the Zahn mechanism can operate very efficiently.

Another interesting result of [63] was the correct value of the mixing depth $\delta M_{\mathrm{mix}} = 0.06 - 0.07$ obtained for the Zahn mechanism, in which δM_{mix} is no longer a free parameter. (Compare this with the semi-empirical value presented above, which was estimated using a diffusion model.) The only remaining free parameter is the mixing rate, which can, in principle, be derived observationally, since D_{mix} is proportional to the square of the rotational velocity in the Zahn mechanism.

An adequate theoretical model explaining the origin of star-to-star variations in elemental abundances in globular clusters should be chosen based on its (at least potential) ability to reproduce *all* available observational data. It turns out that neither mixing scenarios nor primordial nucleosynthesis scenarios are able to do this on their own. The gradual decrease in the C abundance as a star rises along the red giant branch [10], some indications that the Na and Al abundances of giants increase with their luminosity [107, 37], the constancy of the total C+N+O and Mg+Al abundances, and the recent discovery of a very lithium-rich, bright red giant in the globular cluster M3 [104] all provide clear evidence supporting mixing scenarios. However, there are also a whole series of observational facts that are difficult to understand without invoking primordial nucleosynthesis. The clearest example is the variations in the N, Na and O abundances of stars near the main sequence that have been detected in some globular clusters [20, 100, 86].

1.5 A Combined Scenario

It is also difficult to explain the origin of large (> 1 dex) Al overabundances in globular-cluster red giants in pure mixing scenarios.

In our calculations of nucleosynthesis with extra-mixing in globular-cluster red giants, we adopted the following initial chemical composition. First, all the required solar abundances were multiplied by $Z/Z_\odot < 1$, after which the resulting abundances of α elements (in particular, ^{16}O and ^{24}Mg) were increased by a factor of 2.5, in order to achieve agreement with the mean value $\langle [\alpha/\mathrm{Fe}] \rangle \approx +0.4$ observed for Population II field dwarfs [215]. Next, the Na and Al abundances were decreased to [Na/Fe] = [Al/Fe] = −0.4 to take into account the observational fact that, in the presence of low metallicities, the abundances of neutron-rich nuclides are, on average, lower by this amount [215]. Finally, in order to correctly reproduce the global anticorrelation, the ^{22}Ne abundance was also reduced to the value [^{22}Ne/Na] = 0 [66]. This algorithm yields for the Mg isotopic ratios ^{24}Mg/^{25}Mg/^{26}Mg = 90.5/4.5/5.0.

Our calculations indicate that Al cannot form from ^{24}Mg in the stars studied (this requires a temperature of about $75 \cdot 10^6$ K, while the temperatures in red giants do not exceed $55 \cdot 10^6$ K). At the same time, Al can be obtained from ^{25}Mg, but in quantities that are insufficient to explain the observations. The only reasonable solution of this problem is that the initial ^{25}Mg abundances in globular-cluster red giants were enhanced in primordial nucleosynthesis processes (our estimates require that initially [^{25}Mg/Fe] ≥ 1). Thus, we arrive at the need to consider a "combined scenario" taking into account both extra-mixing in red giants and abundance anomalies inherited by these stars, which arose during primordial nucleosynthesis. Then, from the proposed increased contribution of ^{25}Mg to the sum Mg = ^{24}Mg+^{25}Mg+^{26}Mg and the evolutionary transformation of this ^{25}Mg into Al, the total magnesium abundance [Mg/Fe] should appreciably decrease with increasing [Al/Fe]. Such a correlation is, indeed, observed in giants in M13 [169, 107]. However, analysis of the isotopic composition of Mg in a sample of six bright giants in M13 showed that stars with especially high Al abundances have deficits of ^{24}Mg ($\langle[^{24}\text{Mg/Fe}]\rangle = -0.33$ for the five stars with the highest values of [Al/Fe]), and not of ^{25}Mg [170]. This analysis also revealed (for the first time in the entire period of spectral studies of these stars) anomalous Mg isotopic ratios, with an increased contribution of the sum ^{25}Mg+^{26}Mg (+0.21 dex) and mean relative isotopic abundances $\langle^{24}\text{Mg}\rangle/\langle^{25}\text{Mg}\rangle/\langle^{26}\text{Mg}\rangle = 56/22/22$. Unfortunately, spectroscopically the isotopes ^{25}Mg and ^{26}Mg were not separated, and it was therefore assumed that they had the same abundances.

Denissenkov et al. [66] suggested interpreting these observational results using the following combined scenario. They assumed that ^{24}Mg was underabundant for some reason, while the abundance of ^{25}Mg was enhanced in primordial nucleosynthesis in M13. Hydrogen burning and extra-mixing in stars on the red giant branch does not affect the ^{24}Mg abundance, but leads to an increase in the Al abundance due to burning of ^{25}Mg. The sum ^{25}Mg+^{26}Mg increases due to the increase in the abundance of ^{26}Mg, which is also produced from ^{25}Mg. In this case, we eventually obtain approximately the abundances reported by Shetrone [169, 170]. The next problem will be to identify sites of primordial nucleosynthesis in globular clusters in which the required mixture of Mg isotopes could be produced: a deficit of ^{24}Mg and an excess of ^{25}Mg.

1.6 Sites of Primordial Nucleosynthesis in Globular Clusters

Possible external sources of "contamination" of low-mass stars in globular cluster include winds from massive MS stars, supernova explosions, winds and planetary nebulae from AGB stars and, possibly, novae. Are there any signs in globular clusters of nucleosynthesis that occurred in earlier generations of stars? If we suppose that protoclusters formed from material having a cosmological chemical composition with zero metallicity, the answer to this question will be "certainly yes," since a wide range of heavy-element abundances are observed in stars of modern globular clusters.

Massive main-sequence stars and Type II supernovae. It is thought that a massive star with $Z = 0$ loses a negligible amount of its mass in the form of stellar wind [128]. Therefore, the only products of nucleosynthesis in massive first-generation stars that are of interest are nuclides ejected into the interstellar medium during Type II supernovae. This is primarily oxygen (and other α elements). Type Ia supernovae probably do not play an important role in enriching the interstellar medium of globular clusters with heavy elements [111]. We estimated the relative abundances of some NeNa and MgAl nuclides, N and Fe that could arise as a result of multiple Type II supernova explosions and subsequent dilution in the protocluster gas. These are presented in Table 3.1 in Section 3.1. The data in this table show the following:

1. The ^{22}Ne/Na ratio is much lower than the proposed initial ratio for globular-cluster red giants (see above).

2. The abundances of ^{25}Mg and ^{26}Mg are very low compared to those of both ^{24}Mg and Al.

3. There is a large deficit of N.

4. [Fe/H] corresponds to values observed in the most metal-poor globular clusters.

AGB stars. As already noted above, AGB stars are sites of active nucleosynthesis, whose products could enter the interstellar medium either via stellar wind, while the star is still on the AGB, or during the ejection of a planetary nebula, in the final stage of evolution on the AGB. Since intermediate-mass AGB stars ($M \approx 2.5 - 8\,M_\odot$) evolve much more rapidly than low-mass stars, they could play an important role in primordial nucleosynthesis scenarios.

In [66], we studied nucleosynthesis in an AGB star with mass $5\,M_\odot$ and $Z = 10^{-4}$ using a simple parametric model. The density and temperature distributions in regions of nuclear burning, description of convective mixing in the region between the helium and hydrogen burning shell at the time of the helium flash, and parameters describing the convective dredge-up immediately after the flash and mass loss were all taken from exact calculations of stellar evolution on the AGB (for a more detailed description of the computational algorithm, see [68]). Unfortunately, the use of this simple type of method, in which realistic stellar models are replaced with approximate parametric models, is currently the only way of carrying out detailed multi-parameter calculations (for various initial masses and chemical compositions) of nucleosynthesis on the AGB that do not require a colossal expenditure of computer resources. A comparison of the results of nucleosynthesis calculations for intermediate-mass AGB stars obtained using our parametric model with the exact data of [134] obtained using modern evolutionary code for AGB stars shows good agreement. Similar approximation methods were used to investigate nucleosynthesis with extra-mixing in red giants in [64, 43, 161]. Recently, Weiss *et al.* [212] showed that this approach also yields fairly reliable results on the red giant branch.

Table 3.2 in Section 3.1 compares the relative growths (in dex) in the abundances of ^{22}Ne, Na, ^{25}Mg and ^{26}Mg in the region between the helium and hydrogen burning shells in an AGB star with $M = 5\,M_\odot$ after 400 flashes for three initial chemical compositions: (1) solar; (2) $Z = 10^{-4}$ with a relative distribution of heavy elements characteristic of Population II field dwarfs (see above); (3) the composition from Table 3.1. Mixture (3) was obtained in the framework of our primordial nucleosynthesis scenario. Precisely this chemical composition leads to the largest increase in the ^{22}Ne/Na ratio. We can also see that it best reproduces the ^{25}Mg and ^{26}Mg abundances.

1.7 Possible Role of the Cluster Medium

Observations show that extra-mixing is less pronounced in metal-poor field giants than in globular-cluster giants. In particular, there are no field giants with a deficit of O and Mg and an excess of Al, and there are very few field giants with positive values of [Na/Fe] [107]. It is possible that this difference between field and globular-cluster giants is associated entirely with characteristics of the formation and evolution of globular clusters (in short, with the role of the cluster medium). In a two-part combined scenario, the cluster medium could influence both the part associated with primordial nucleosynthesis and that associated with extra-mixing. Indeed, globular clusters are such dense stellar aggregates that processes "contaminating" low-mass stars (ejecta from supernovae and AGB stars) could play a more substantial role in them than in the galactic field. For example, it is possible that, due to the high energy of the explosion, some material ejected by a supernova becomes gravitationally unbound from the protocluster and leaves it. On the contrary, material lost by intermediate-mass AGB stars (via both stellar wind and the ejection of planetary nebulae) has an appreciably lower kinetic energy, and is likely to remain in the cluster. Consequently, stars of the next generation in the globular cluster can form from material with a higher proportion of nuclides synthesised in AGB stars, but with the same [Fe/H] value (since iron-group elements are synthesised only in superovae). The role of close binary systems in the production of anomalous stellar surface abundances could also be more important in globular clusters than in field stars, since the relative number of such binaries could be appreciably higher in the dense cores of clusters. On the other hand, extra-mixing generated by rotation (or by some other mechanism) might operate more effectively in globular-cluster red giants than in field giants (for some as yet unknown reason).

1.8 Structure of the Review

The review consists of the Introduction (Section 1), three sections presenting the main results of our studies on subjects closely related to the problems of abundance anomalies in globular clusters, the Conclusion (Section 5) and a list of references.

Section 2 is dedicated to the problem of deep mixing in globular-cluster red giants. Its first subsection describes a diffusion model for extra-mixing, which is then used to explain the global anticorrelation of [O/Fe] with [Na/Fe] and the correlation of [C/Fe] with M_V for giants in M92. Here, we note the unfortunate fact that, during our work on [64] (in 1994–1995), there were no sufficiently reliable simultaneous determinations of surface abundances for several elements involved in the CNO, NeNa, and MgAl cycles in giants within a single globular cluster. In particular, the samples of M92 giants used to obtain the O–Na anticorrelation and the correlation of [C/Fe] with M_V were non-overlapping. This circumstance was the basis for our choice of various sets of parameters for the diffusion mixing (the mixing depth and rate) in our attempts to theoretically reproduce these observational correlations. Jumping ahead for a moment, we note that publications of observational results for four (anti)correlations between the abundances of various elements (C–N, C–O, O–Na and O–Al) for giants within a single cluster appeared slightly later: in 1995 for ω Cen [139] and in 1999 for M4 [96].

In Section 2.2, we propose Zahn's mechanism [219, 132] for extra-mixing in red giants, which considers the joint action of meridional circulation and turbulent diffusion generated by the star's rotation. It is shown that, with the rotational velocity required to provide the observed mixing rates in globular-cluster red giants, the mixing depth for the Zahn mechanism automatically coincides with the semi-empirical value obtained for the diffusion model.

In Section 2.3, we propose a solution to the problem of anomalously high lithium abundances observed in a small number of red giants. We consider the recent discovery of a lithium-rich giant in M3 [104]. The production of a sufficiently large quantity of ^{7}Li in a red giant model with extra-mixing requires a mixing rate that is two to three orders of magnitude higher than is required by the global anticorrelation of [O/Fe] with [Na/Fe]. In the context of the Zahn mechanism, this means that the rotational velocity of a lithium-rich red giant must be at least an order of magnitude higher than the mean rotational velocity for ordinary giants, and should approach the local Keplerian velocity. We suggest the absorption of giant planets as a mechanism for the spontaneous acceleration of the rotation of lithium-rich giants.

All the work considered in the review has been carried out over the past seven years. It essentially reflects the development of ongoing scientific discussions between the proponents of two alternative types of scenarios initially proposed more than two decades ago to explain the origin of star-to-star variations in elemental abundances in globular clusters: primordial nucleosynthesis scenarios and mixing scenarios. Ideas about the actual relative contributions of these two mechanisms have changed with time, depending on *(i)* the standard rates for NeNa- and MgAl-cycle reactions accepted at a given time and *(ii)* the available observational data on the abundances of Na, CNO-cycle and MgAl-cycle elements in globular-cluster stars, in particular, in those that are close to the main sequence. In connection with this, we indicate at the beginning of each section of the review the year (or years) in which work on the corresponding publications was carried out.

In our paper [66], discussed in Section 3.1, we first proposed a symbiosis of the two scenarios indicated above – a "combined scenario." This was dictated by the appearance of new observational data – the discovery of variations in the abundances of C, N, O, Na, Al and Mg in stars near the main sequences of several globular clusters (see, for example, [189, 20, 31, 47, 100, 86]). These variations could only be the result of accretion by these stars of small amounts of material that have undergone nuclear processing in the presence of hydrostatic hydrogen burning in stars of previous generations. The most suitable candidates for such stars "contaminating" the interstellar medium of globular clusters are AGB stars of low mass (suppliers of *s*-process elements) and intermediate mass (suppliers of products of the *s*-process and hydrogen burning at the base of the convective envelope) (see Section 3.1), as well as red giants with masses $0.9 \leq M/M_\odot \leq 2.5$ in which extra-mixing has occurred (Section 3.2). All these stars have already finished their evolution (leaving behind white dwarfs) in modern globular clusters.

One of the central aspects of the combined scenario is the production of appreciable quantities of the isotope ^{25}Mg in intermediate-mass, low-metallicity AGB stars. An initial excess of ^{25}Mg in low-mass stars inherited from the interstellar medium is required to explain the production of Al in globular-cluster red giants. Unfortunately, remaining uncertainties in the rates of MgAl-cycle reactions admit two alternative interpretations for Al overabundances in globular-cluster red giants: either these overabundances are purely ^{27}Al (Section 3.1) or they are nearly entirely made up of the radioactive isotope ^{26}Al^g (corresponding to the modification of the combined scenario discussed in Section 3.2).

Based on the combined scenario, we have proposed a model for the chemical evolution of globular clusters, which is described in the Conclusion.

Finally, in Section 4, we consider the transition of MS stars with masses of 10 and 30 $M_\odot$ to a state of stationary rotation. We assume that evolutionary variations in the profiles of the specific angular momentum and mixing rate in the radiative envelope of a massive star are due to the joint action of meridional circulation and turbulent diffusion (the Zahn mechanism), as in the radiative zone of a red giant. This analysis is important, since it addresses the question of why there is an appreciable delay (of up to 50% of the main-sequence lifetime) in the appearance of products of the CNO cycle at the surfaces of massive MS stars, which, as a rule, rotate rapidly, while, on the contrary, extra-mixing apparently operates rather efficiently in red giants. An answer is provided by differences in the behaviour of material in the radiative zones of these two types of stars. The material in the radiative zone of a massive MS star is under nearly steady-state conditions. The action of meridional circulation and turbulent diffusion compensate each other. As a result, mixing becomes very slow, and a rather long time is required for the "wave" of diffusion to cover the distance from the region of hydrogen burning at the boundary of the convective core to the stellar surface. In the radiative zone of a red giant, there is a constant inward flow of hydrogen-rich material feeding the hydrogen burning shell. As a result of conservation of specific angular momentum, the rotation of this material must speed up as it moves inward. Consequently, if there were no extra-mixing in the radiative zone of the red giant, this would support a state of strongly differential rotation. Our calculations show that extra-mixing in the context of the Zahn mechanism exerts very little influence on the rotational state in the red giant, which, in turn, facilitates efficient mixing. Thus, it may be that the same mixing mechanism operates in different types of stars.

2. Extra-Mixing in Red Giants

2.1 Diffusion Model for Extra-Mixing in Globular-Cluster Red Giants (1994–1995)

2.1.1 Introduction

It has been known for a long time that the abundances of C and N vary from star to star in globular clusters (see the reviews by Smith [173], Suntzeff [187, 188] and Kraft [103]). These variations are reflected in the intensities of the CN, CH and NH molecular bands: weak CH bands testify to low C abundances, while strong CN and NH bands indicate N excesses. The most surprising and as yet unexplained property of the distribution of CN-band intensities for globular-cluster stars is its bimodality for clusters with metallicities in the interval $-1.6 \leq [Fe/H] \leq -0.7$. Groups of stars with weak and strong CN bands are separated by a rather distinct gap in the histogram of CN-band intensities [175, 18]. It is interesting that this bimodality is observed in globular clusters with moderately low metallicities, such as M4, M5 and NGC 6752, with $[Fe/H] = -1.2$, -1.17 and -1.3 respectively [183, 189], but is not apparent in clusters with very low metallicities, such as M15 and M92, with $[Fe/H] = -2.3$ and -2.25 [182].

We should emphasise that, apart from the abundances of C and N, only the abundances of O, Na and Al vary appreciably from star to star in globular clusters, while the abundances of other elements remain virtually the same (within the margin of error of 0.2 dex) within a single cluster (exceptions are ω Cen and M22) [165, 103]. The most important circumstance here is undoubtedly the observational fact that the abundances of these elements are correlated with each other, which very probably indicates that their corresponding variations arose in a single place. Usually, stars with excess abundances of N also possess enhanced abundances of Na and Al, low

DEEP MIXING IN GLOBULAR-CLUSTER RED GIANTS   23

isotopic ratios $^{12}C/^{13}C$, and deficits of C, sometimes with deficits of O as well, compared to "normal" stars. As an example of such correlations, Fig. 2.1a presents the anticorrelation of [O/Fe] with [Na/Fe] for giants in M3 and M13 ([Fe/H] $= -1.47$ and -1.49) and for field giants ([Fe/H] from -2.8 to -1.2) according to the data of Kraft *et al.* [106]. The obvious increase in the Na abundance as [O/Fe] decreases visible in Fig. 2.1a is actually a general property displayed by more than 70 giants in various globular clusters that have been studied up to the present time. For this reason, Kraft *et al.* [106] called this observational relation the "global anticorrelation" of [O/Fe] with [Na/Fe].

Certain characteristic features visible in Fig. 2.1a require some commentary. First, there is a plateau in the observed dependence of [Na/Fe] on [O/Fe]. This may indicate a depletion of the nuclear source of Na. Second, while the relation for the cluster M13 is fairly well populated in the region of low O abundances ($-0.75 \le$ [Fe/H] ≤ -0.25), M3 only has giants with [O/Fe] ≥ -0.2, while the studied field giants have [O/Fe] ≥ 0.0. Third, a rather large variation in [Na/Fe] is observed for stars with high oxygen abundances ([O/Fe] ≥ 0.0). All these observational properties may carry useful information for theories devised to explain the origin of the global anticorrelation.

Another, less striking, example is the correlation of the isotopic ratio $^{12}C/^{13}C$ with [C/Fe] (Fig. 2.1b). The form of the dependence of $^{12}C/^{13}C$ on [C/Fe] differs strongly from cluster to cluster. A number of globular clusters show fairly large variations in atmospheric oxygen abundance for a single value of $^{12}C/^{13}C$. Often, the observed values of $^{12}C/^{13}C$ are close to the equilibrium value for the CNO cycle (≈ 3.5), appreciably lower than is predicted by standard stellar evolution theory. Moreover, Sneden *et al.* [185] have shown that the isotopic ratio $^{12}C/^{13}C$ in metal-poor field stars apparently decreases with M_V, which could be an indication of evolution. Jumping ahead for a moment, we note that our computations show that $^{12}C/^{13}C$ is very sensitive to extra-mixing parameters in red giants.

According to standard concepts, the stars under consideration are completing their first ascent along the red giant branch. They consist of an isothermal, degenerate helium core surrounded by a hydrogen burning shell and an extended convective envelope. Between the hydrogen burning shell and the base of the convective envelope is a

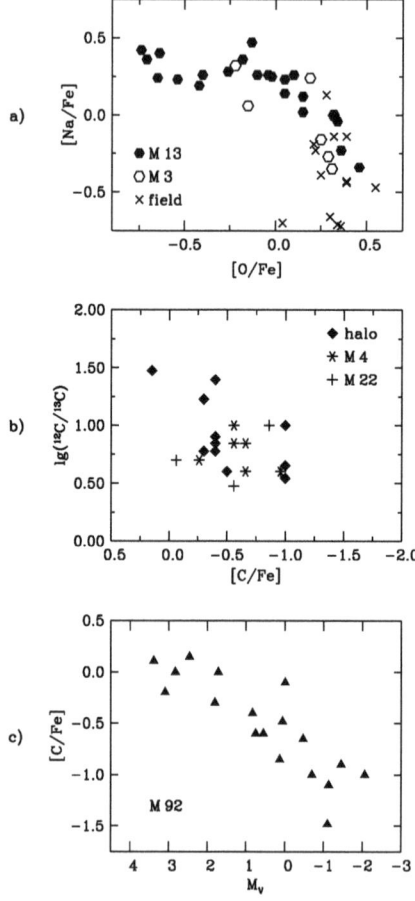

Figure 2.1. a) Global anticorrelation of [O/Fe] with [Na/Fe] for field and globular-cluster red giants [106]. **b)** Correlation of $^{12}\text{C}/^{13}\text{C}$ with [C/Fe] for Population II giants (observational data are taken from [185] for halo stars and from [176] for M4 and M22; the vertical axis plots $\lg(^{12}\text{C}/^{13}\text{C})$). **c)** Correlation of [C/Fe] with absolute V magnitude for stars of the metal-poor cluster M92. Here, we use only the data of Langer *et al.* [117] not marked by a colon in that work, indicating them to be relatively poorly-determined values with small weights. We also omit data for AGB stars.

DEEP MIXING IN GLOBULAR-CLUSTER RED GIANTS

zone with moderate mass that is in radiative equilibrium. Depending on its luminosity, the star either has or has not already passed through the first dredge-up, when the base of the convective envelope reaches its greatest depth, after which it begins to recede outward. Standard evolutionary calculations show that the variations in the atmospheric abundances of elements and their isotopes occurring in the first dredge-up in low-metallicity stars are negligible, and cannot explain the anomalies observed in globular-cluster red giants. For example, after the first dredge-up in model red giants with masses of $M = 0.8$ and 1.0, the surface isotopic ratio $^{12}C/^{13}C$ which depends strongly on both the metallicity and mass of the star, falls to values of 67 (in our computations, 65) and 29 for $Z = 10^{-4}$, and to 50 and 28 for $Z = 10^{-3}$, respectively [185]. The standard theory does not predict variations in the surface abundances of O, Na and Al on the red giant branch.

Two hypotheses were put forward to explain the characteristic chemical compositions of globular-cluster stars: (1) abundance anomalies were inherited from the products of primordial nucleosynthesis in globular clusters, and the different elemental abundances in different stars within a single cluster were already present in the material from which the currently observed stars were formed; (2) variations in stellar atmospheric abundances are due to the action of some extra-mixing that carries the products of nuclear reactions from layers adjacent to the hydrogen burning shell to the base of the convective envelope. There are observational facts and theoretical arguments supporting both of these hypotheses. The bimodality in the distribution of CN-band intensities has traditionally been taken as evidence for inherited abundance anomalies. This bimodality can be traced clearly in the clusters NGC 6752, M5 and 47 Tuc, right to the time when the stars leave the main sequence [189, 18]; that is the bimodal behaviour was already present in the stars before the onset of extra-mixing on the red giant branch (see below). This provided weighty evidence supporting the primordial nucleosynthesis hypothesis, since the numerical simulations of VandenBerg and Smith [203] showed that extra-mixing in evolutionary stages before and on the main sequence is not able to explain the observed bimodality in the CN-intensity distribution.

On the other hand, a decisive observational fact supporting extra-mixing scenarios is the correlation of [C/Fe] with absolute magnitude found in a number of clusters, such as M3, M13 and M15 (see [187] and references therein). The most pronounced correlation between [C/Fe] and M_V is observed for the metal-poor cluster M92 (see Fig. 2.1c, for which the data of Langer *et al.* [117] were used). This correlation directly demonstrates that the atmospheric abundance of carbon decreases as stars rise along the red giant branch. Note that a decrease in the C abundance with an increase in the $(B-V)$ colour index (i.e. essentially with an increase in luminosity) has been found even in the giant clusters M4 and NGC 6752, in which bimodal distributions of the CN-band intensity are observed [189].

These facts lead us to conclude that the star-to-star variations in elemental abundances observed in globular clusters are due to the action of two independent factors: inhomogeneities in the chemical composition of the protocluster and the transport of products of nuclear burning to the atmosphere of the red giant [103].

The first theoretical model for a red giant with extra-mixing was proposed by Sweigart and Mengel [192], who considered rotation-induced meridional circulation as a means of transporting material reprocessed in the CN cycle from the hydrogen burning shell to the base of the convective envelope. They also proposed an explanation for the fact that deep extra-mixing is encountered primarily in metal-poor giants. In their analysis of the chemical structures of model red-giants, Sweigart and Mengel discovered that, at low metallicities, the layer in which C is transformed into N in CN-cycle reactions (the "C layer") is quite distinct from the hydrogen burning shell, so that the hydrogen abundance in this layer remains nearly equal to the atmospheric abundance. Consequently, if there are meridional circulation flows below the base of the convective envelope, they can easily reach the C layer, where the molecular-weight gradient is too low to stop them. In the presence of even lower metallicities, even the O layer (the place with the highest temperature, where O is transformed into N in ON-cycle reactions) is higher than the bulk of the hydrogen burning shell.

In the 1970s and 1980s, proponents of the primordial nucleosynthesis hypothesis were encouraged by the discovery of Cohen [46] and Cottrell and Da Costa [49] that globular-cluster stars with strong CN

DEEP MIXING IN GLOBULAR-CLUSTER RED GIANTS 27

bands also display Na and Al excesses. At that time, it was not at all understood how Na and Al could be synthesised in stars whose energy source was hydrostatic hydrogen burning. However, in 1987, Denissenkov and Ivanov [62] demonstrated that the resonance reaction ^{22}Ne(p,γ)^{23}Na produces a significant quantity of Na during hydrogen burning in the convective cores of intermediate- and high-mass MS stars (see also the note by Kudryashov and Tutukov [109], who discussed the synthesis of Na and Al during hydrogen burning in intermediate-mass AGB stars). Slightly later, Denissenkov and Denissenkova [60] established that if [^{22}Ne/Na] = 0 the same reaction is capable of producing the necessary amount of Na in the O layers of evolving red giants. This last result explained the Na–N correlation, which was already known at that time, and anticipated the Na–O anticorrelation. In their subsequent work, Denissenkov and Denissenkova [61] computed the theoretical dependence of the atmospheric abundance [O/Fe] on [Na/Fe] for red giants. They assumed that the extra-mixing acting in the zone between the hydrogen burning shell and the base of the convective envelope was sufficiently rapid to support a uniform chemical composition in this zone, but was slower than the convection in the envelope. It is striking that the global anticorrelation of [O/Fe] with [Na/Fe] discovered later (Fig. 2.1a) precisely follows the theoretical dependence proposed by Denissenkov and Denissenkova.

In 1993, Langer et al. [115] considered nucleosynthesis near the hydrogen burning shell in a metal-poor red giant in more detail than in [61]. In particular, they traced variations in the abundances of 85 isotopes, from hydrogen to calcium, and used a more realistic initial chemical composition. In addition to the results obtained in [61], Langer et al. found that Al could also be synthesised in the O layer in the reaction ^{26}Mg(p,γ)^{27}Al, and that even more Na could be produced due to the much more abundant isotope of neon ^{20}Ne.

At the beginning of our work on [64], only two models with extra-mixing in the radiative zone of a red giant that could be used to trace the *evolution* of elemental abundances at the stellar surface were being discussed in the literature.

VandenBerg and Smith [203] artificially placed the base of the convective envelope near the C layer. The true depth of this "expanded" convective envelope was chosen by hand such that the C abundance at the surface decreased by a specified small relative amount (0.005) between successive models. VandenBerg and Smith successfully reproduced the initial part (to $M_V \approx +0.3$) of the decrease in [C/Fe] in the cluster M92 (Fig. 2.1c).

Three other simple models were analysed in [61]: the extra-mixing in the radiative zone was assumed to be (1) "slow" (not exerting an appreciable influence on the abundance profiles); (2) "rapid" (capable of maintaining a uniform chemical composition in the radiative zone, but still acting much more slowly than convective mixing in the envelope) and (3) "very rapid" (with a rate comparable to that of convective mixing). It was shown that the rapid-mixing case was in closest agreement with the observations; the correlation between the abundances of CNO elements and Na discovered in 1989 by Paltoglou and Norris [142] for giants in ω Cen is quite satisfactorily reproduced. It was also noted that an even closer agreement with the observations could be obtained for extra-mixing at a rate between "rapid" and "slow," but this required verification using a corresponding intermediate model, such as a diffusion model.

In this section, we are concerned with the following problem. We consider an evolutionary model with masses and metallicities characteristic of the stars in metal-poor globular clusters from the base to the tip of the red giant branch. We assume that there is some extra-mixing in the radiative zone, between the hydrogen burning shell and the base of the convective envelope, which we model using diffusion, with the depth and rate (diffusion coefficient) of this mixing treated as free parameters. The main question is whether it is possible to find reasonable values for these parameters that reproduce all three correlations depicted in Figs. 2.1a, 2.1b and 2.1c.

2.1.2 Models and Computation Method

As stellar model parameters representative of the current population of metal-poor globular clusters, we chose a mass of $M = 0.8\,M_\odot$ and relative mass abundances of heavy elements and helium of $Z = 10^{-4}$ and $Y = 0.22$ respectively. We calculated the evolution of such a model using the code described in [151]. The input physics of this evolutionary code includes the OPAL table of opacity coefficients from [157], supplemented by tables from [213] at low temperatures.

The energy released in pp-chain and CNO-cycle reactions was computed using the reaction rates of [36] (it is important to note that we used a finer nuclear-reaction grid when calculating the nucleosynthesis kinetics taking into account diffusion; see below). We did not allow for mass loss due to stellar wind, since this is negligible for low-mass, metal-poor stars on the red giant branch. We likewise did not consider convective overshooting or semi-convection. We described ordinary convection using standard mixing-length theory, with the ratio of the mixing length to the pressure scale height equal to 1.55, as is required for the construction of solar models using this code. Extra-mixing was not taken into account in the evolutionary computations, which were carried out in order to derive basis models for red giants (see below).

Our basis models of red giants were four models for stars rising along the red giant branch, selected from an evolutionary sequence and used as reference "points" for interpolation in our calculations for nucleosynthesis with extra-mixing (see Fig. 2.2a, for which we used the bolometric correction of VandenBerg [202] to translate the theoretical stellar luminosities into M_V values). The first of the selected basis models (No. 610 in the evolutionary sequence) is located significantly lower than the point corresponding to the end of the first dredge-up (marked by the solid square), and the second (No. 1010 in the evolutionary sequence) is located just beyond this point. The solid triangle in Fig. 2.2a marks the point at which the outward-moving hydrogen burning shell crosses the jump in the hydrogen abundance distribution left by the base of the convective envelope at the moment of its deepest penetration into the star. In the terminology of Gratton *et al.* [87], the "upper" red-giant branch begins at this point. Sweigart and Mengel [192] presented the following theoretical arguments supporting the hypothesis that extra-mixing in red giants begins precisely at this point. In order for the meridional-circulation flows to influence the abundances of CNO elements in the envelope, two conditions must be fulfilled: (1) there should not be too high a molecular-weight gradient in the radiative zone separating the convective envelope from the hydrogen burning shell, and (2) the temperature at the base of the region of circulation flows should be high enough for the CN-cycle reactions to take place there. Neither of these conditions can be fulfilled until the hydrogen burning shell reaches the jump in the distribution of the H abundance. This is confirmed by observational data for field and open-cluster giants [84, 42, 43, 45, 87]. However, the situation with very metal-poor giants is apparently special, since observations indicate that, in some cases, extra-mixing is able to begin acting much earlier (Fig. 2.1c). Accordingly, we carried out calculations for nucleosynthesis with extra-mixing beginning both with model 1010 (case "A") and with model 610.

Since the mass of the helium core M_c gradually increases as the star ascends the red giant branch, it is convenient to introduce the relative mass co-ordinate $\delta M = (M_r - M_c)/(M_{\rm BCE} - M_c)$, where M_r is the usual Lagrangian co-ordinate, defined as the mass enclosed inside a sphere of radius r, and $M_{\rm BCE}$ is the value of M_r at the base of the convective envelope. In fact, the mass of the hydrogen burning shell is negligibly small compared to M_c, so we can always assume that $M_{\rm HBS} \approx M_c$ independent of the precise meaning assigned to the Lagrangian co-ordinate of the hydrogen burning shell $M_{\rm HBS}$ (the co-ordinate at its base, middle or upper edge). As M_r varies from M_c (or $M_{\rm HBS}$) to $M_{\rm BCE}$, δM takes on values from 0 to 1. To calculate the abundance profiles in the radiative zone, we must know the temperature, density and radius as functions of M_r for each time step. We used the following procedure for this purpose.

The growth in the mass of the helium core over a time Δt is roughly equal to

$$\Delta M_c \approx \frac{L}{X_{\rm CE} E_{\rm H}} \Delta t,$$

where L and $X_{\rm CE}$ are the current luminosity of the star and the relative hydrogen mass abundance in the convective envelope; $E_{\rm H} \approx 6.3 \cdot 10^{18}$ erg·g^{-1} is the energy released in the burning of one gram of hydrogen. Consequently, we can estimate the mass of the helium core at any time as

$$M_c(t) \approx M_c(t_{n_i}) + \int_{t_{n_i}}^{t} \frac{L}{X_{\rm CE} E_{\rm H}} \, dt,$$

where t_{n_i} is the age of the initial model ($n_i = 610$ or 1010). Having estimated M_c in this way, we can find $\Delta M = M_{\rm BCE} - M_c$ and L at

DEEP MIXING IN GLOBULAR-CLUSTER RED GIANTS

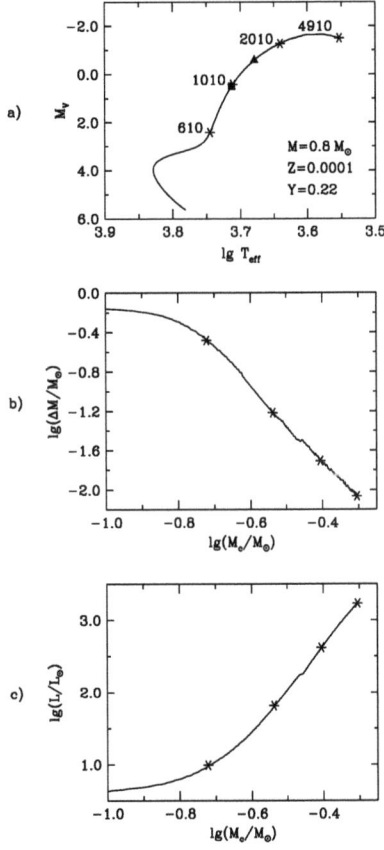

Figure 2.2. a) Evolutionary track for a stellar model with mass and metal content typical of red giants of metal-poor globular clusters. The asterisks and corresponding numbers denote the four models used when interpolating the distributions of structural parameters (temperature, density, etc.). The end of the first dredge-up is marked by a square, and a triangle marks the model in which the hydrogen burning shell crosses the jump in hydrogen abundance left by the base of the convective envelope at the time of its deepest penetration into the star. b) and c) illustrate how the dependences of both $\lg(\Delta M/M_\odot)$ and $\lg(L/L_\odot)$ on the logarithm of helium core mass can be approximated as linear functions.

an arbitrary time t, using approximation relations of the form

$$\lg(\Delta M/M_\odot) = \lg k_1 + a_1 \lg(M_c/M_\odot)$$

$$\lg(L/L_\odot) = \lg k_2 + a_2 \lg(M_c/M_\odot),$$

whose validity is illustrated in Figs. 2.2b and 2.2c. Further, the values of $\lg T(t, \delta M)$ can be obtained via interpolation between the known distributions $\lg T(t_{n_i}, \delta M)$ and $\lg T(t_{n_{i+1}}, \delta M)$ (Fig. 2.3a), where $t_{n_i} \leq t \leq t_{n_{i+1}}$ ($n_i = 610$, 1010 or 2010). After this, we transform the relative mass co-ordinate δM back into the usual Lagrangian co-ordinate $M_r = M_c(t) + \Delta M \cdot \delta M$ and, finally, obtain $\lg T(t, M_r)$ for $M_c(t) \leq M_r \leq M_{\rm BCE}(t)$. Together with the new hydrogen abundance in the envelope $X_{\rm CE}(t)$, obtained from the solution of the nuclear kinetics equations taking into account diffusion (see below), the estimated value of $L(t)$ is used to continue this procedure.

In this section, we model extra-mixing with diffusion without indicating or discussing the associated physical mechanism. The evolution of the abundance profiles for 24 nuclides taking part in 30 nuclear reactions in pp chains and the CNO, NeNa and MgAl cycles are traced by solving a system of nuclear kinetics equations (with allowance for diffusion) in the form

$$\frac{\partial y_i}{\partial t} = \left(\frac{\partial y_i}{\partial t}\right)_{\rm nucl} + \frac{\partial}{\partial M_r}\left[(4\pi r^2 \rho)^2 D_{\rm mix} \frac{\partial y_i}{\partial M_r}\right], \quad i = 1, \ldots, 24, \quad (2.1)$$

where $y_i = N_i/(\rho N_A)$, N_i is the number density of the ith nuclide and N_A is Avogadro's number. The corresponding nuclear reaction rates were taken from [36]. Eq. (2.1) is supplemented by the two boundary conditions

$$\frac{\partial y_i}{\partial M_r} = 0 \quad M_r = M_c$$

and

$$\frac{\partial y_i}{\partial t} = -\frac{(4\pi r^2 \rho)^2 D_{\rm mix}}{(M - M_{\rm BCE})}\frac{\partial y_i}{\partial M_r} \quad M_r = M_{\rm BCE}, \quad (2.2)$$

where M is the mass of the star.

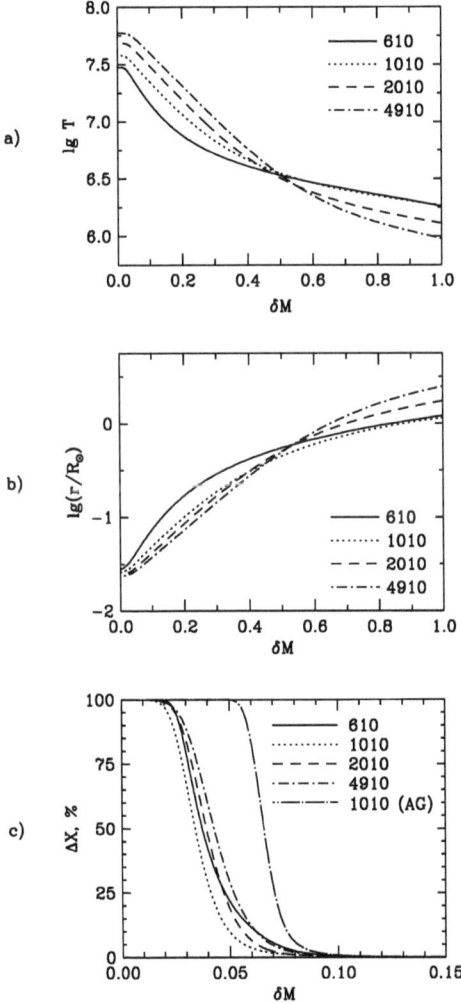

Figure 2.3. a) and b) lg T and lg($r/R_\odot$) as functions of the relative mass coordinate δM (defined in the text) for the four basis models; c) Amount of hydrogen (in percent) burned at a depth δM in stellar models without extra mixing. The curve No. 1010 (AG) was calculated using model No. 1010, but with a solar initial chemical composition [2].

We introduce two parameters: the diffusion coefficient D_mix determining the rate of the extra-mixing and the relative mass coordinate δM_mix specifying the position of the lower boundary of that part of the radiative zone in which extra-mixing occurs (the upper boundary is always taken to coincide with M_BCE). We will assume that $D_\mathrm{mix} = 0$ when $\delta M < \delta M_\mathrm{mix}$.

To solve the system of partial differential equations (2.1–2.2), we replaced the time and mass derivatives appearing in them with corresponding difference ratios, and approximated the nuclide abundances on the right-hand sides with sums of the form $y_i(k) + \Delta_i(k)$, where $\Delta_i(k)$ is the variation in the abundance of nuclide i at the point with mass co-ordinate M_k over a time Δt. The quadratic terms $\Delta_i(k)\Delta_j(k)$ were initially set equal to zero, and the resulting system of linear algebraic equations was solved for the corrections $\Delta_i(k)$ using the Gauss elimination method taking into account the sparseness of the coefficient matrices. In the next iteration step, the non-linear terms $\Delta_i(k)\Delta_j(k)$ were replaced by the products of the corresponding corrections found in the previous step, and the system of equations was solved again. This procedure was repeated until the required accuracy was achieved. As a rule, only three iterations were required for the magnitudes of the differences of the ratios $\Delta_i(k)/y_i$ obtained in two successive iterations to become less than 10^{-4} at all mass points. This algorithm, which is actually a generalisation of the method used by Maeder [127], makes it possible to trace variations in the abundance distributions within the star with fairly high accuracy, even in the presence of diffusion. Test computations showed that the total number of nucleons $\sum_i A_i y_i$ (A_i is the atomic mass number) which should be conserved within a star over its evolution, remained constant in our calculations within an absolute error of 10^{-8}. The characteristic computational time step was 10^4–10^5 years.

Modelling extra-mixing using basis models for red giants, as described above, has the advantage that the entire domain of admissible values for the free parameters can be investigated easily and rapidly. In addition, it is possible to use a fairly large nuclear reaction grid. The main drawback of this procedure is that we cannot feed back into the algorithm the influence of diffusion on the evolution of the star. We assume that this influence is negligible. It is obvious that this assumption is valid only as long as extra-mixing does not affect the region of the hydrogen burning shell, which produces a substantial amount of nuclear energy, and as long as the chemical composition of the envelope (more precisely, its hydrogen content) is not subject to appreciable variations.

With regard to the first of these conditions, our calculations show that, even in the most extreme case we investigated, more than 99% of the luminosity of the star is produced below the depth considered by us, δM_{mix}.

With regard to the second condition, variations in X_{CE} shift the star away from standard evolutionary tracks in a colour–magnitude diagram, which can be observed as a spreading of the red giant branch. VandenBerg and Smith [203] found that extra-mixing leading to a dependence of [C/Fe] on M_V similar to that observed in the cluster M92 decreases X_{CE} by only about 4%, consistent with the width of the red giant branch in M92. The mixing depths used by us lowered the atmospheric H abundance by no more than 5–10% (the only exception is the special case discussed in the last paragraph of Section 2.1.3).

We essentially used the same initial chemical composition as Langer et al. [115]. Namely, initially "solar" abundances of all isotopes heavier than ^{4}He were lowered by a factor of $Z_\odot/Z \approx 0.019/10^{-4} = 190$ (using the data of [2]); further, the abundances of the "α isotopes" ^{16}O, ^{20}Ne, ^{24}Mg and ^{28}Si were increased by a factor of 2.5, since observations indicate that $[\alpha/\mathrm{Fe}] \approx +0.4$ in metal-poor field and globular-cluster giants (see, for example, [215]). For the initial abundance of ^{22}Ne, we used the two values $[^{22}\mathrm{Ne/Na}] = 0$ and $[^{22}\mathrm{Ne/Na}] = 0.4$, with $[\mathrm{Na/Fe}] = -0.4$.

What values of the free parameters D_{mix} and δM_{mix} should be considered "reasonable", that is, admissible? If there are large-scale flows in the red giant (not unlike meridional circulation) that mix material in the radiative zone, then, to first approximation, the diffusion coefficient can be calculated as the product $D_{\mathrm{mix}} \sim vl$, where v and l are the characteristic velocity and linear scale of the flows. As a theoretical estimate of the rate of meridional circulation in red giants, Sweigart and Mengel [192] obtained the value $v_{\mathrm{circ}} \sim 10^{-3}$ cm·s^{-1}. They used the rotation rates of low-mass MS stars as their input observational data, and considered two simple laws for the distribution of the specific angular momentum in the giant's envelope.

Smith and Tout [177] attempted to derive a semi-empirical estimate of $v_{\rm circ}$ using a simple analytical model that reproduced the correlation between [C/Fe] and M_V in M92 (Fig. 2.1c). They concluded that $v_{\rm circ} \sim 10^{-2}$ cm·s^{-1}. We can adopt the difference $r_{\rm BCE} - r_{\rm c}$ for l, which is of order $\sim 10^{11}$ cm (Fig. 2.3b). Thus, reasonable estimates for the rate of extra-mixing are $D_{\rm mix} \sim 10^8 - 10^9$ cm^2·s^{-1}.

We preferred to work with $\delta M_{\rm mix}$ rather than $M_r^{\rm mix}$, so that the mixing depth could be considered a fixed parameter. The latter quantity increases with time due to the outward motion of the mass of the hydrogen burning shell. In fact, maintaining a constant value of $\delta M_{\rm mix}$ means that, at the lower boundary of the region with extra-mixing, the temperature must gradually increase with the star's luminosity, since the function $\lg T(t, \delta M)$ becomes steeper with time (Fig. 2.3a). Of course both quantitiies – $D_{\rm mix}$ and $\delta M_{\rm mix}$ – can change during the evolution of a real star. Unfortunately, we do not yet understand how this occurs. The only obvious restriction on $\delta M_{\rm mix}$ is that it cannot take on values that are very small, say $\delta M_{\rm mix} < 0.04$ (see below). Otherwise, the extra-mixing would approach the hydrogen burning shell too closely and exert an appreciable influence on the star's evolution on the red giant branch, which is not observed [203]. Figure 2.3c plots the amount of hydrogen ΔX (in percent) that is burned at a specified depth δM as a function of this depth for our four basis models. We can see that reasonable values for the mixing depth are $\delta M_{\rm mix} > 0.06 - 0.07$. In this case, $\Delta X \leq 5 - 10\%$ at the base of the region with extra-mixing.

2.1.3 Results

Some characteristic properties of the abundance profiles in models without mixing Figure 2.4 presents abundance profiles for some isotopes near the hydrogen burning shell in two red giant models with which we initiated our diffusion computations (we have used the relative mass abundances $X_i = y_i A_i$, where A_i is the atomic mass number of the isotope).

The differences in the depicted profiles for models 610 and 1010 are primarily due to their different temperature distributions. For the range of δM values shown in Fig. 2.4, the temperature of model 1010 is somewhat higher (see Fig. 2.3a), giving rise to the small

DEEP MIXING IN GLOBULAR-CLUSTER RED GIANTS

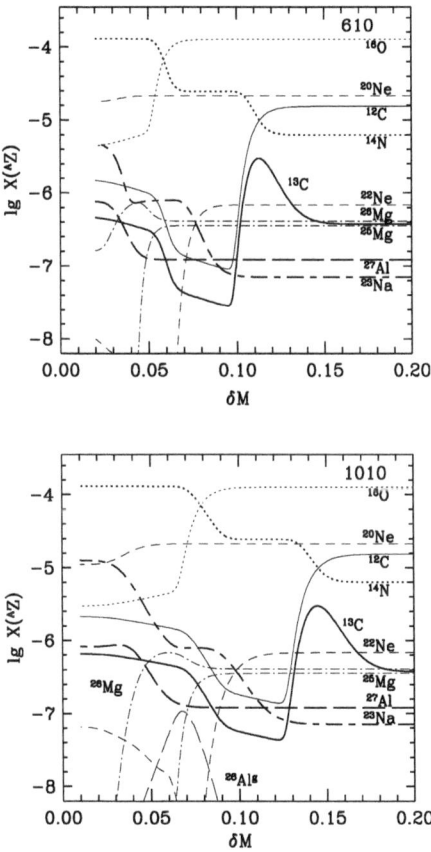

Figure 2.4. Logarithm of relative (mass) abundances X_i of some isotopes as a function of the relative mass co-ordinate δM in two initial models for the diffusion-mixing calculations. The model numbers are indicated in the upper right corner of each panel.

outward shift of the profiles and other differences. For example, the Na-abundance profile has two distinct steps, the right-hand one due to the synthesis of Na from ^{22}Ne in the reaction ^{22}Ne(p,γ)^{23}Na and the left-hand one associated with the partial transformation of ^{20}Ne into Na. Due to the higher temperature in model 1010, a slightly larger amount of ^{20}Ne forms Na, making the left step in the abundance profile higher than in model 610, but leaving the height of the right step unchanged, since it is determined by the initial abundance of ^{22}Ne, which was the same in the two models.

Another interesting feature that merits attention is the increase in the abundance of ^{27}Al due to burning of ^{25,26}Mg directly below the O layer. This was first pointed out by Langer et al. [115], who, however, did not establish the true origin of this. In fact, both of these properties for the red giant model with low metallicity – the appearance of a second (left) step in the Na-abundance profile and an increase in the ^{27}Al abundance – are brought about by the low value of Z, not the higher temperature below the O layer. To demonstrate this, we illustrate in Fig. 2.5 abundance profiles for these same isotopes (excluding only ^{12}C and ^{13}C) *for the same distributions of temperature and density* as in the lower panel of Fig. 2.4, but assuming that the initial chemical composition was solar (using the values from [2]). Figure 2.5 clearly shows a total absence of the second step in the Na-abundance profile (i.e. there is now no transformation of ^{20}Ne into Na) and of any production of ^{27}Al. These differences can be interpreted as follows.

Let us consider the nuclear kinetics equation for isotope i. It can have a form such as

$$\frac{dy_i}{dt} = \lambda(T)\rho y_j X \pm \ldots, \qquad (2.3)$$

where X is the hydrogen abundance and $\lambda(T)$ is the rate of the reaction involving isotope j and hydrogen that leads to the formation of isotope i. We have $Z \ll Z_\odot$ and a distribution of the initial abundances of isotopes heavier than helium in the proportion $y_i/Z = y_i^\odot/Z_\odot$. If (2.3) is divided by Z, we obtain a kinetics equation in a form that is invariant with respect to Z

$$\frac{d(y_i/Z)}{dt} = \lambda(T)\rho(y_j/Z)X \pm \ldots. \qquad (2.4)$$

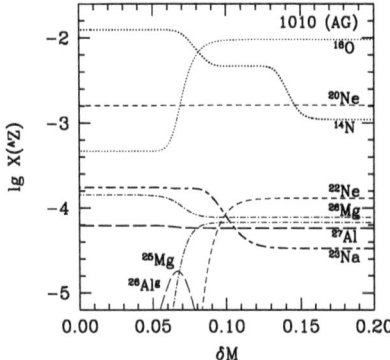

Figure 2.5. The abundance profiles of the same isotopes (excluding ^{12}C and ^{13}C) as in Fig. 2.4 in model 1010, but computed assuming that the intial chemical composition of the star was solar.

If all the kinetics equations used in our calculations had this form, the solutions for various Z values would be similar, and could be derived from the solution for some specific value Z' using the proportion relation $y_i = (Z/Z')y'_i$.

However, the equation for hydrogen has a different form:

$$\frac{dX}{dt} = [-\lambda(T)\rho(y_j/Z)X \mp \ldots] \cdot Z. \tag{2.5}$$

It follows that choosing a small Z leads to a decrease in the hydrogen burning rate; i.e. to an increase in the time over which the abundances of other nuclides can vary. This can be expressed in other words as follows. At low Z values, the total amount of CNO elements is also low, so that for the same T and ρ, the CNO cycle requires more time to burn all the hydrogen. At lower Z values, the NeNa cycle has more time to operate, and this time proves to be sufficient for the abundance of ^{20}Ne to begin to decrease, and for the second step in the Na-abundance profile to emerge. The same is true of the production of ^{27}Al in the case of low Z.

Evolution of the surface abundances Figure 2.6 shows the results of our computations that best reproduce the correlation between [C/Fe] and M_V in M92. The initial model used in all these computatons was No. 610. We can see that diffusion mixing must be rather slow ($D_{\text{mix}} \approx 5 \cdot 10^7 - 10^8$ cm$^2 \cdot$s^{-1}) and shallow ($\delta M_{\text{mix}} \approx 0.09 - 0.10$) if it is to reproduce the observed decrease in the C abundance with M_V (Fig. 2.6c).

The following four features in Fig. 2.6 are also worthy of attention.

1. It turns out that the theoretical dependence of [C/Fe] on M_V is very sensitive to the choice of both mixing parameters (Fig. 2.6c), so that the fact that the corresponding observed correlation appears fairly distinct (even taking into account the possible scatter in the initial C abundances) means that the dominant parameter determining the extra-mixing in globular-cluster stars does not vary strongly from star to star, at least not in M92.

2. All the theoretical curves in Fig. 2.6c have a plateau at low values of M_V. Consequently, the AGB stars in M92 should also have underabundances of C, close to those observed for the red giants. Indeed, this is apparently the case [32].

3. The calculated isotopic ratio ^{12}C/^{13}C drops extremely rapidly at the very beginning of the extra-mixing (Fig. 2.6b). This is in agreement with the observational data of Suntzeff and Smith [189], who found that red giants in the globular clusters M4 and NGC 6752, whose stars have bimodal distributions of the CN-band intensity, have very low ratios ^{12}C/^{13}C (sometimes even equal to the equilibrium value), independent of their observed [C/Fe] values, although these latter values occupy a broad interval from -0.25 to -1.12.

4. We conclude from Fig. 2.6a that only a modest amount of Na can be carried to the surface of a star displaying the dependence of [C/Fe] on M_V shown in Fig. 2.6c, while the surface abundance of O should not change. The global anticorrelation of [O/Fe] with [Na/Fe] established by Kraft et al. [106] includes a sample of ten red giants in M92, which follow this dependence all the

DEEP MIXING IN GLOBULAR-CLUSTER RED GIANTS 41

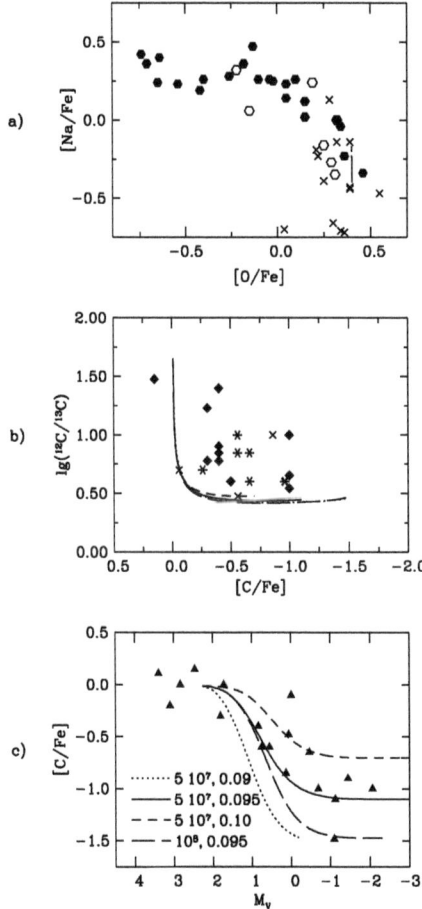

Figure 2.6. Results of calculations for deep diffusion mixing that best reproduce the correlation of [C/Fe] with M_V observed for M92. The mixing rate and depth are indicated in panel **c**.

way to [O/Fe] ≈ 0.0. Unfortunately, this sample does not overlap with the sample of stars that demonstrate the correlation between [C/Fe] and M_V.

Our conclusions 1 and 3 are in agreement with those of Vanden-Berg and Smith [203]. We should add that the calculation results presented here are qualitatively the same as those obtained when the mass of the star is changed from $0.8\,M_\odot$ to $0.9\,M_\odot$.

Figure 2.7 presents the results of our calculations with diffusion mixing that best reproduce the global anticorrelation of [O/Fe] with [Na/Fe]. The solid curve in Fig. 2.7a was computed assuming that $[^{22}\text{Ne/Na}] = 0.4$. We can see that too much Na is produced in this case. If we suppose that the global anticorrelation is, indeed, due to the reaction $^{22}\text{Ne}(p,\gamma)^{23}\text{Na}$ occurring near the O layer and to deep extra-mixing in red giants, we must assume that initially $[^{22}\text{Ne/Na}] \approx 0.0$ in globular cluster stars. The initial Na abundance does not affect this assumption as long as it remains much lower than the initial ^{22}Ne abundance. Other calculations in which we used the initial ratio $[^{22}\text{Ne/Na}] = 0.0$ are marked with a **B** (recall that **A** denotes computations using model 1010 rather than model 610 as the initial model).

The diffusion mixing must be rather rapid ($D_{\text{mix}} \approx 5 \cdot 10^9$ cm$^2 \cdot$s^{-1}) and **deep** ($\delta M_{\text{mix}} \approx 0.06-0.07$) if it is to explain the observed growth of the surface abundance of Na as [O/Fe] decreases (Fig. 2.7a). However, these mixing rates and depths remain "reasonable" in the sense discussed in Sect. 2.1.2. We will now summarise other important aspects of the results of computations with rapid deep diffusion mixing.

1. The chosen mixing depth guarantees that only the first (right) step in the Na-abundance profile will be affected by the mixing, and will take part in the production of the excess Na at the giant's surface (see Fig. 2.4). If we chose a lower value of δM_{mix}, say 0.05, the second (left) step would also be influenced by the mixing, in which case we would obtain a much higher surface Na excess, in contradiction with the observations, even if we assume $[^{22}\text{Ne/Na}] = 0.0$. However, we have already emphasised that choosing δM_{mix} values lower than 0.06–0.07 could lead to undesirable variations in the overall behaviour of the star's

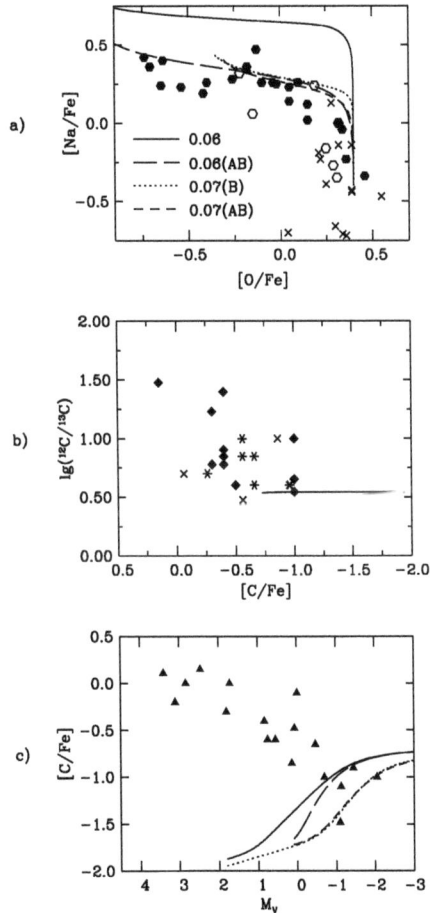

Figure 2.7. Diffusion calculations that reproduce the global anticorrelation of [O/Fe] with [Na/Fe]. They were all carried out using the same value $D_{\text{mix}} = 5 \cdot 10^9$ cm$^2 \cdot$s^{-1}. The mixing depths are indicated in panel **a**. Calculations using model 1010 rather than model 610 as an initial model are marked with an **A**. Calculations carried out assuming that [^{22}Ne/Na]= 0.0 rather than 0.4 are marked with a **B**.

evolution on the red giant branch, due to the transport of too much He to its atmosphere.

2. The results of calculations beginning with models 1010 (**A**) and 610 are virtually indistinguishable in Fig. 2.7.

3. Due to the high mixing rate, [C/Fe] falls to very low values near −2.0 at the very beginning of the calculations, after which it gradually increases to ≈ -0.75. This growth is due to the increase in the N abundance brought about by the formation of N from O in the ON cycle, since, under equilibrium conditions, the C abundance varies in proportion to the N abundance. The isotopic ratio $^{12}C/^{13}C$ at the surface nearly always remains close to its equilibrium value. If such rapid mixing existed in real red giants, we would observe a correlation between [C/Fe] and M_V with the opposite sign than that observed for the cluster M92 (Fig. 2.7c). Unfortunately, in both cases considered – slow, shallow and rapid, deep mixing – the final C abundances are roughly the same. Therefore, we cannot distinguish which of these two regimes more closely corresponds to the real situation by determining [C/Fe] for stars near the tip of the red giant branch. The resolution of this problem may prove to be even more difficult if the properties of mixing in real stars vary with time.

4. In our calculations, the ^{27}Al abundance increased by a factor of approximately three. However, this is not sufficient to explain the observational data. For example, red giants with unusually high excess abundances of Al (by factors of up to 20!) have been detected in the clusters M13 [168] and ω Cen [139]; these stars also display overabundances of Na and underabundances of O (see the solid hexagons in Fig. 2.7a). To explain these data, we must either assume that initially $[^{25,26}Mg/Al] \approx 0.9$ rather than 0.4 in stars in these globular clusters, or search for sources of primordial Al synthesis in them.

The correlation between $^{12}C/^{13}C$ and [C/Fe] is less distinct than the other two correlations depicted in Fig. 2.1 (we cannot even rule out the possibility that there is no correlation). In the diffusion calculations for which results are discussed above, the equilibrium ratios

DEEP MIXING IN GLOBULAR-CLUSTER RED GIANTS   45

^{12}C/^{13}C at a giant surface are established almost immediately after "turning on" the extra-mixing, and it is tempting to think that this might be generally true. However, it turns out that it is possible to chose mixing parameters that lead to a correlation between ^{12}C/^{13}C and [C/Fe], if we consider a monotonic, almost linear decrease in the logarithm of ^{12}C/^{13}C with [C/Fe] a correlation. Fig. 2.8 presents the results of such calculations, all of which begin with model 1010 (**A**) and used a mixing depth of $\delta M_{\mathrm{mix}} = 0.04$; i.e. they can all be characterised as calculations with "very deep mixing". It is evident that this value for δM_{mix} cannot be considered "reasonable". Therefore, we take the corresponding results to reflect an extreme case that is unlikely to be realised in globular cluster giants.

1. In this case, the second step in the Na-abundance profile also contributes to the production of the surface Na excess, which accordingly becomes very large (recall that we continue to use the initial ratio [^{22}Ne/Na] = 0.0). The more rapid the mixing, the closer the global anticorrelation comes to the corresponding theoretical curve (Fig. 2.8a), however, at the same time, the resulting values of [O/Fe] are too low.

2. Very deep mixing yields an almost linear dependence of the logarithm of the ratio ^{12}C/^{13}C as a function of [C/Fe] (Fig. 2.8b), since it maintains a nearly constant ^{13}C abundance at the surface of the giant. To understand why the ^{13}C abundance remains constant, let us consider Fig. 2.4. We can see that the ^{13}C profile has a "hump" next to a "dip". Very deep mixing encompasses both the hump and the dip, so that the ^{13}C excess in the hump is compensated by the deficit in the dip.

3. The surface abundance of ^{12}C is not very low, since almost immediately after the onset of the mixing, a near-equilibrium state is established in the CNO cycle (even in the ratios of surface abundances), and [C/Fe] varies in proportion to the N abundance, which has grown due to O burning (Fig. 2.8c). Indeed, if we examine the lower panel in Fig. 2.4, we can see that, for a depth $\delta M_{\mathrm{mix}} = 0.04$, the extra-mixing penetrates to the region where the entire CNO cycle operates in equilibrium. At greater depths, the N abundance doubles, now due to O and

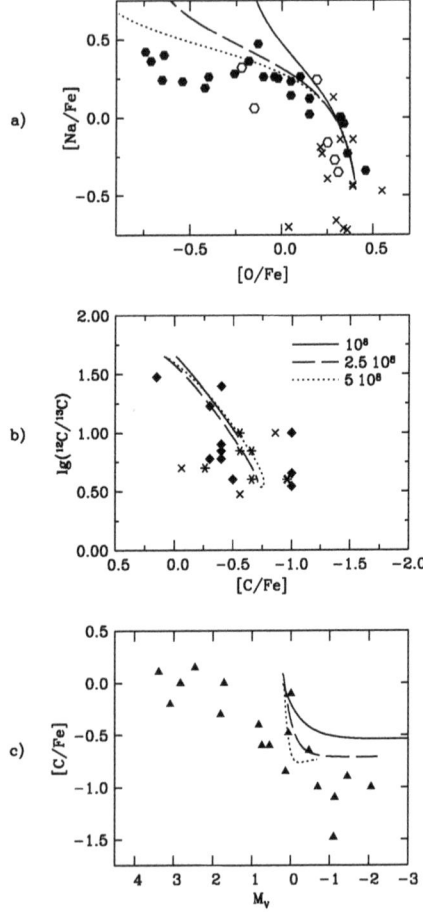

Figure 2.8. Results of diffusion calculations that reproduce the correlation between $\lg(^{12}C/^{13}C)$ and [C/Fe]. The initial model was No. 1010 (**A**), and the mixing depth was taken to be $\delta M_{\mathrm{mix}} = 0.04$. The diffusion coefficients used are indicated in panel **b**.

not C burning, and ^{12}C and ^{13}C behave in the same fashion as N. However, at $\delta M_{\rm mix} = 0.04$ and deeper, the abundances of all CNO elements cease to change, since they have already achieved their equilibrium values there. The plateau in the [C/Fe] profile at low values of $M_{\rm V}$ in Figs. 2.7c and 2.8c is due precisely to the surface ^{12}C abundance achieving its equilibrium value.

Dilution of material undergoing nuclear reprocessing in the convective hydrogen core The hypothesis of primordial nucleosynthesis serves as an alternative to deep mixing. Two initial processes have been proposed to explain the O–Na anticorrelation and N–Na correlation. Cottrell and Da Costa [49] have suggested that the s-process in flashes of the helium burning shell in intermediate-mass AGB stars could have led to an enhancement of the interstellar media of globular clusters in nitrogen, sodium and aluminium in the past. Another possible mechanism is hydrogen burning in massive MS stars, accompanied by intense mass loss due to their stellar winds, which could "contaminate" the interstellar medium with products of the CNO, NeNa and MgAl cycles (Freeman [77]). Here, we present dependences of [Na/Fe] on [O/Fe] that could result from hydrogen burning in the convective core of a massive, metal-poor MS star. We used a very simple model, in which we assume that the logarithm of the temperature $\lg T(M_r)$ in the convective core, in which there is instantaneous mixing, decreases in a linear fashion from the centre of the star, where $T_{\rm c} = 3 \cdot 10^7$ K, to the surface of the core, where $T_{\rm s} = 1.5 \cdot 10^7$ K, and that the density is constant throughout and equal to $\rho = 1$ g·cm^{-3}. The initial abundances of nuclides were taken to be the same as in our models for globular-cluster red giants with [^{22}Ne/Na] $= 0.0$ (case **B**). We traced the nucleosynthesis in the convective core as long as the hydrogen abundance in this core $X_{\rm c}$ exceeded a specified value. Further, the reprocessed core material was artificially mixed with material having the initial chemical composition. The results of such dilution are shown in Fig. 2.9.

To reproduce the global anticorrelation in a primordial nucleosynthesis scenario with hydrogen burning in the convective cores of massive stars, we must suppose that we observe a mixture of material in the atmospheres of globular-cluster red giants in which only 1–5% of the hydrogen has burned. This type of mixture can easily be created

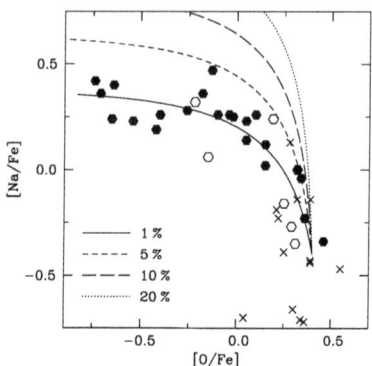

Figure 2.9. Dependence of [Na/Fe] on [O/Fe] in the case of dilution of material undergoing partial nuclear reprocessing in a metal-poor, convective hydrogen core. The percentage of hydrogen burned in the core up to the moment of (artificial) mixing with unprocessed material is indicated in the bottom left-hand corner.

in an evolving red giant simply by enabling extra-mixing to penetrate to the depth where $\Delta X = 1 - 5\%$ (Fig. 2.3c), but it is difficult to imagine how such a mixture could come about via the dilution of material that has undergone only 5% of its nuclear reprocessing in the convective core of a massive MS star.

Deep diffusion mixing and the relative abundances of oxygen isotopes In 1994, Boothroyd et al. [15] pointed out that it is impossible to explain certain observational data on the $^{17}O/^{16}O$ and $^{18}O/^{16}O$ isotopic ratios for low-mass ($M \leq 1.7\,M_\odot$) AGB stars of spectral types S and C with approximately solar metallicities, and also accurate measurements of O isotopic abundances in meteorites, as the result of standard convective dredge-up or hydrogen burning at the bases of the convective envelopes in intermediate-mass AGB stars. This "forbidden" region of O isotopic ratios is delineated by the dashed line in Fig. 2.10. Boothroyd et al. [16] proposed that a process similar to extra-mixing in globular-cluster red giants may be responsible for the fact that some low-mass AGB stars fall within this region.

DEEP MIXING IN GLOBULAR-CLUSTER RED GIANTS 49

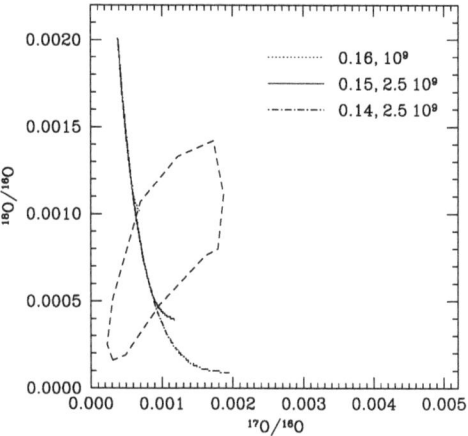

Figure 2.10. Deep diffusion mixing on the red giant branch in a star with $M = 1.2\,M_\odot$ and solar metallicity leads to surface isotopic ratios for oxygen which are in agreement with those observed in S and C stars on the AGB. The region occupied by these stars is delineated by the dashed line (data taken from Boothroyd et al. [16]); it is not possible to explain the presence of stars in this region using standard stellar evolution theory. The mixing rate and depth used in our calculations are indicated in the top right-hand corner.

Signs of extra-mixing such as a decrease in the ratio of the ^{12}C and ^{13}C abundances with increasing stellar luminosity have also been detected in red giants with solar metallicities [84, 42]. Therefore, we tried to move stars into the delineated region in Fig. 2.10 using the same diffusion-mixing model that we used earlier to interpret the star-to-star surface abundance variations in globular clusters. However, we now used giant models with $M = 1.2\,M_\odot$ and $Z = 0.02$ as the basis, and adopted the rates for the reactions ^{17}O(p,α)^{14}N and ^{17}O(p,γ)^{18}F published by Landré et al. [112] and the uncertainty coefficients $f_1 = 0.2$ and $f_2 = 0.1$ (case N1 in the notation of Boothroyd et al. [16]). The initial abundances of elements heavier than helium were taken to be solar, using the values of [2]. We did not correct the initial isotopic abundances of oxygen to take into account variations occurring in the first dredge-up, since these are negligible compared to variations produced by extra-mixing on the red giant branch. We assumed that extra-mixing begins when the hydrogen burning shell crosses the jump in the hydrogen abundance left by the base of the convective envelope when it reaches its maximum size at the end of the first dredge-up. The validity of this assumption is indirectly confirmed by the observed surface abundances in red giants with approximately solar metallicity [42]. The calculation results are presented in Fig. 2.10. We can see that if extra-mixing is present in red giants, the O isotopic ratios could well reach the region forbidden by standard stellar evolution theory. The mixing depth and rate implied in this case are quite "reasonable" (they are presented in Fig. 2.10).

2.1.4 Main Conclusions

We have shown that deep diffusion mixing can explain all three observed correlations in Fig. 2.1. However, the results of our computations can reproduce each of these correlations using different mixing depths and rates. Unfortunately, the observational correlations were obtained for different groups of stars. In particular, the sample of stars in the cluster M92 for which the correlation between [C/Fe] and M_V was derived (Fig. 2.1c) and the sample of M92 stars for which the global anticorrelation of [O/Fe] with [Na/Fe] is traced [106] do not overlap. We have established that the theoretical dependences resulting from our attempts to reproduce the correlations in Fig. 2.1 are very sensitive to the choice of mixing parameters. This all makes the search for correlations for stars within a single sample such as those in Fig. 2.1 an urgent observational task, which can serve as an effective test for deep-mixing scenarios.

2.2 A Possible Mechanism for Extra-Mixing in Globular-Cluster Red Giants (1999)

2.2.1 Introduction

The search for a mechanism for extra-mixing in globular-cluster red giants remains an unsolved problem of stellar astrophysics. It appears that this mixing differs from ordinary convection. It is thought that it acts in the radiative zone separating the hydrogen burning

DEEP MIXING IN GLOBULAR-CLUSTER RED GIANTS 51

shell and the base of the convective envelope in stars ascending along the red giant branch. Extra-mixing is manifested by variations in the atmospheric abundances of isotopes of C, N, O, Na, Mg and Al when the mixing is able to penetrate sufficiently close to the hydrogen burning shell, where these isotopes take part in the transformation of H into He and act as catalysts in reactions in the CNO, NeNa and MgAl cycles. Indeed, observations have revealed variations in the star-to-star atmospheric abundances of these elements in globular clusters, and also in the ratio ^{12}C/^{13}C (see the numerous references in [66]). Moreover, there are observational correlations between the abundances of C, N, O, Na, Mg and Al that are in agreement with theoretical expectations. For example, the abundances of the pairs C and N, O and Na, and Mg and Al are anticorrelated. This is illustrated in Fig. 2.11, which shows the abundance distributions of some nuclides near the hydrogen burning shell as functions of the relative mass co-ordinate $\delta M = (M_r - M_c)/(M_{bce} - M_c)$. Here, the usual Lagrangian mass co-ordinate M_r is the mass contained within a sphere of radius r, M_c is the mass of the helium core (more precisely, the Lagrangian co-ordinate of the point at which the relative hydrogen content is $X = 10^{-4}$) and, finally, M_{bce} is the co-ordinate of the base of the convective envelope. When composing Fig. 2.11, we used the same programme to compute the nucleosynthesis kinetics, and the red giant model with luminosity $\lg L/L_\odot = 2.08$ from the same evolutionary sequence ($M = 0.8\,M_\odot$, $Z = 5 \cdot 10^{-4}$) as in [66]. The appreciable increase in the ^{27}Al abundance with depth is due to our application of non-standard assumptions; namely, we assumed an enhanced initial abundance of the isotope ^{25}Mg ([^{25}Mg/Fe] = 1.1) and a rate of the reaction ^{26}Al^g(p,γ)^{27}Si that was higher than that published in [5] by a factor of 10^3.

The so-called aluminium problem is that it is not possible to explain the large Al excesses observed for red giants (such as those seen for giants in the globular cluster ω Cen in Fig. 2.13b) by invoking extra-mixing alone. In a mixture of isotopes, ^{24}Mg usually dominates over ^{25}Mg and ^{26}Mg: the Mg isotopic ratios in the solar system are ^{24}Mg:^{25}Mg:^{26}Mg=79:10:11. As we can see in Fig. 2.11, the ^{27}Al in a red giant cannot be produced from ^{24}Mg, since this requires temperatures higher than $60 \cdot 10^6$ K, while the temperature in the hydrogen burning shell does not exceed $55 \cdot 10^6$ K. On the

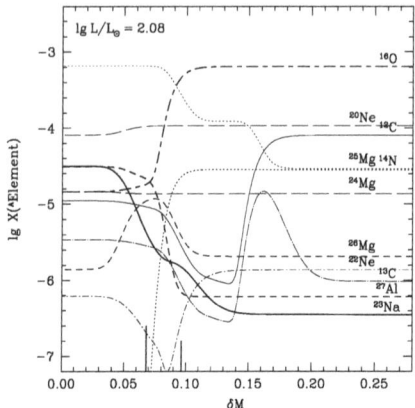

Figure 2.11. Abundances ($X_i = A_i y_i$, where A_i is the atomic mass number and y_i is defined in equation (2.6)) of some nuclides participating in reactions of the CNO, NeNa and MgAl cycles as functions of a relative mass co-ordinate ($\delta M = 0$ at the base of the hydrogen burning shell and $\delta M = 1$ at the lower boundary of the convective envelope) in the red giant model with which we began our nucleosynthesis computations taking into account extra-mixing. The two vertical line segments on the horizontal axis mark places where $D_v = 0$ (the smaller segment) and $U = 0$ (larger segment) (see text).

other hand, ^{25}Mg can be transformed into Al, but the resulting Al abundance is not high enough to explain the observations if initially [^{25}Mg/Fe]=0. The only reasonable solution to this problem is the hypothesis that the ^{25}Mg abundances in some globular-cluster red giants were initially enhanced as a consequence of previous chemical evolution of the clusters. It was shown in [66] that previous-generation intermediate-mass AGB stars could produce the required variations in the isotopic ratios of magnesium in globular clusters. However, in this case, ^{25}Mg must efficiently form ^{27}Al in the reactions ^{25}Mg(p,γ)^{26}Al^g(p,γ)^{27}Si($\beta^+\nu$)^{27}Al, even in the presence of a fairly high abundance of ^{25}Mg. It turns out that this does not happen when the reaction rates from [36] are used: the competing chain of reactions ^{25}Mg(p,γ)^{26}Al^{g}($\beta^+\nu$)^{26}Mg operates more rapidly. As a result, a large fraction of the ^{25}Mg forms ^{26}Mg, and not ^{27}Al. The role of this undesired effect can be reduced by increasing the rate of the reaction ^{26}Al^{g}(p,γ)^{27}Si. According to [5], this reaction rate may be underestimated by a factor of 10^3 in the range of temperatures characteristic of the hydrogen burning shells of red giants ($T \approx 40-50 \cdot 10^6$ K). This is precisely why we increased the rate of this reaction by a comparable factor in [66]. Using this increased reaction rate and assuming that initially [^{25}Mg/Fe]=1.1, we were able to reproduce quite satisfactorily the anticorrelation of [O/Fe] with [Al/Fe] for giants in ω Cen with our theoretical relation (Fig. 2.13b, dashed line). The rates of MgAl-cycle reactions were recently examined in [4, 3]. The upper limit for the rate of the reaction ^{26}Al^{g}(p,γ)^{27}Si now exceeds the value computed using the data of [36] by a factor of $10^{1.5}$ only. In connection with this, we have proposed an alternative solution to the aluminium problem, which is discussed in Section 3.2.

If we examine Fig. 2.11 and imagine that the extra-mixing reaches layers with, say, $\delta M = \delta M_{\text{mix}} \approx 0.06$, we can immediately conclude that such mixing should enrich the envelope of the giant in N, Na and Al and make it poorer in C, O and ^{25}Mg. The hypothesis that Na was synthesised near the hydrogen burning shell and carried outward in globular-cluster red giants was put forth and justified by Denissenkov and Denissenkova in 1990 [61]. Subsequently (in 1993), support for this hypothesis was found by Langer *et al.* [115], and it was supplemented with the idea that Al was synthesised together with Na. Denissenkov and Denissenkova also essentially predicted the anticorrelation between O and Na in globular-cluster red giants that was discovered later in [105]. As a consequence of the fact that this anticorrelation is observed in nearly all globular clusters that have been studied at the current time, Kraft [103] referred to this as the "global anticorrelation" of [O/Fe] with [Na/Fe]. This anticorrelation can clearly be seen in Fig. 2.12, where the solid hexagons indicate the most extreme case, observed for M13 [107]. A distinguishing characteristic of this last case is the presence of giants that are extremely poor in oxygen ([O/Fe]≈ -0.8).

Based only on the observed correlations between the abundances of different elements, we cannot exclude the possibility that they have

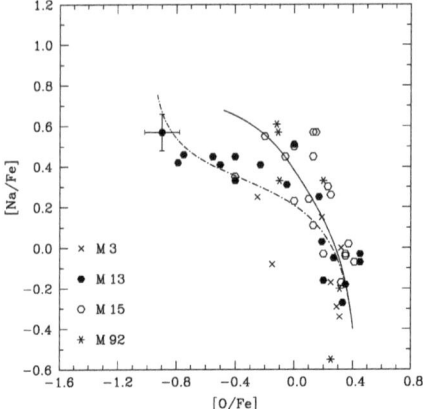

Figure 2.12. Global anticorrelation of [O/Fe] with [Na/Fe] constructed for four globular clusters and theoretical relations predicted by a semi-empirical diffusion model (dot–dashed line; the mixing depth and rate are $\delta M_{\mathrm{mix}} = 0.06$ and $D_{\mathrm{mix}} = 2.5 \cdot 10^9$ cm$^2 \cdot$s^{-1}, respectively) and a Zahn model (solid line; see text for details). The observational data for M3 are from [105], for M13 from [107], for M15 from [184] and for M92 from [169]. Typical observational errors are shown for M13.

come about as a result of primordial nucleosynthesis in these globular clusters; i.e. in processes that varied the chemical composition of the cluster material before the formation of the currently observed stars. However, an evolutionary scenario in which these correlations are associated with nuclear reactions and deep extra-mixing in the evolved red giants themselves is supported by the observed decrease in [C/Fe] with M_V (i.e. with an increase in stellar luminosity) detected in a number of globular clusters [9, 117, 17, 189]. There is also evidence for a gradual increase in the Na abundance and a decrease in the O abundance with increasing luminosity of giants in M13 [107].

Sweigart [191] has discussed the problem of extra-mixing in globular-cluster red giants in connection with the old problem of the so-called "second parameter", which, together with metallicity – the "first parameter" – determines the morphology of the horizontal branch in a globular cluster. Sweigart speculated that, if extra-mixing was sufficiently deep and rapid, it could appreciably enhance the He abundance in the atmospheres of red giants, which, in turn, would lead to the formation of very blue horizontal-branch stars after the core helium flash.

Some extra-mixing is also present in intermediate- and high-mass MS stars. Its action can explain the excesses of N and He observed in the atmospheres of OB stars [91, 126]. Such mixing has also been invoked to solve the so-called mass discrepancy problem for OB stars [113, 211, 59]. It is known that massive MS stars rotate rapidly. Therefore, it is natural to suppose that extra-mixing in their interiors is somehow connected with their rotation. The most promising mechanism for the mixing associated with rotation was proposed by Zahn [219], and later developed further by Maeder and Zahn [132]. In this mechanism, extra-mixing is produced by the joint action of meridional circulation and turbulent diffusion. Recently, Talon et al. [196] and Denissenkov et al. [67] demonstrated that the Zahn mechanism was potentially capable of explaining the abundance anomalies observed in massive MS stars, and Talon and Charbonnel [194] successfully applied this mechanism to interpret the origin of the hot side of the "lithium gap" in the Hyades.

The main goal of this section is to demonstrate that the Zahn mechanism may also be responsible for star-to-star variations in the elemental abundances of globular-cluster red giants. This is supported observationally by the fact that the most rapidly rotating blue horizontal-branch stars are observed in M13, which displays an extremely low O abundance [145]. Charbonnel [43] was the first to point out the potentially important role of the Zahn mechanism in red giants, but she did not obtain a self-consistent solution to the entire system of Zahn equations. Instead, she simply used the expression for the diffusion coefficient presented by Zahn to model the Li abundances and $^{12}C/^{13}C$ ratios in Population II giants.

2.2.2 Semi-Empirical Diffusion Model for Extra-Mixing

Denissenkov and Weiss [64] and Denissenkov et al. [66] (see also Section 2.1) modelled extra-mixing in red giants using diffusion, without specifying the corresponding physical mechanism. They used a rather simple algorithm. Three red giant models were chosen from a standard evolutionary sequence with metallicity $Z = 5 \cdot 10^{-4}$ in [66]. The first model[3] had luminosity $\lg L/L_\odot = 2.08$, which, in model red giants with $M = 0.8\,M_\odot$, is approximately the luminosity at which the hydrogen burning shell crosses the jump in the hydrogen abundance distribution formed by the base of the convective envelope at the moment of its deepest penetration into the star in the first dredge-up. In spite of our incomplete understanding of the physical mechanism for extra-mixing, it is widely believed that it cannot overcome a barrier created by a sufficiently high gradient of the mean molecular weight [45], but, unfortunately, the precise value for the critical gradient remains unknown. The chemical structure of our initial model – with an absence of jumps in the hydrogen abundance distribution in the radiative zone between the point δM_{mix} near the hydrogen burning shell and the base of the convective zone – guaranteed that extra-mixing would not be suppressed in this zone. The third model selected was located near the tip of the red giant branch, while the second had intermediate luminosity. These three "basis" models were used to interpolate the temperature T, density ρ and radius r in luminosity (essentially, in time) and in the mass co-ordinate δM.

We then solved a system of nuclear kinetics equations in the form

$$\frac{\partial y_i}{\partial t} = \left(\frac{\partial y_i}{\partial t}\right)_{\mathrm{nucl}} + \frac{\partial}{\partial M_r}\left[(4\pi r^2 \rho)^2 D_{\mathrm{mix}} \frac{\partial y_i}{\partial M_r}\right], \qquad (2.6)$$

where $y_i = N_i/(\rho N_A)$, N_i is the number density of nuclide i and N_A is Avogadro's number. The term $(\partial y_i/\partial t)_{\mathrm{nucl}}$ takes into account all nuclear reactions in which nuclide i is produced or destroyed, and D_{mix} is the constant diffusion coefficient. Thus, two parameters were uesd in [64] and [66] to describe the extra-mixing: its depth δM_{mix} and rate D_{mix}. The main task undertaken in [66] was to show that the entire spectrum of star-to-star abundance variations in an individual globular cluster could be reproduced theoretically with the same set of mixing parameters.

The globular clusters ω Cen and M13 were chosen for a comparison of the results with the observations. Although there is a large

---

[3] Used to compile Fig. 2.11.

internal scatter in the metallicities of the stars in ω Cen, $-1.8 \leq$ [Fe/H] ≤ -0.8, distinguishing ω Cen from most other clusters in this sense, it provides a unique opportunity to compare the theoretical predictions simultaneously for four different observational correlations between abundances in red giants (Fig. 2.13). In addition, ω Cen has one of the bluest horizontal branches among globular clusters [216], which, by analogy with M13, may be a consequence of rapid internal rotation of its red giants. A fairly uniformly and densely populated global anticorrelation of [O/Fe] with [Na/Fe] is observed for M13, which extends to the region of very low O abundances (Fig. 2.12). It is fortuitous that both the metallicity of M13 ([Fe/H] $= -1.49$) and the mean metallicity of stars in ω Cen can be fairly well approximated using the same value $Z = 5 \cdot 10^{-4}$ ($\lg Z/Z_\odot = -1.58$).

It was found in [66] that the global anticorrelation in M13 can be interpreted as the result of diffusion extra-mixing with $\delta M_{\rm mix} = 0.06$ and $D_{\rm mix} = 2.5 \cdot 10^9$ cm$^2 \cdot$s^{-1} (Fig. 2.12, dot–dashed curve). At the same time, all four correlations in ω Cen were reproduced sufficiently well by a diffusion model with mixing depth and rate $\delta M_{\rm mix} = 0.05$ and $D_{\rm mix} = 5 \cdot 10^8$ cm$^2 \cdot$s^{-1} (Fig. 2.13, dashed curves).

Analysis of the results obtained using the diffusion model described led us to the following important conclusions. Let us return to the global anticorrelation between [O/Fe] and [Na/Fe] (Fig. 2.12) and suppose that it has a purely (or at least predominantly) evolutionary origin. In that case, its extent in the vertical direction (i.e. the scatter of [Na/Fe] within the anticorrelation) is primarily due to the depth of the mixing. Namely, this scatter indicates that $\delta M_{\rm mix}$ cannot be less than about 0.06–0.07. Indeed, extra-mixing with $\delta M_{\rm mix} \leq 0.06$ would affect the second step in the Na-abundance profile (if it moves inward), whose presence is due to an additional production of Na from ^{20}Ne (over and above that for the channel ^{22}Ne(p,γ)^{23}Na) (Fig. 2.11). Being an α element, the isotope ^{20}Ne is much more abundant than the neutron-enriched isotope ^{22}Ne. The burning of ^{22}Ne forms the first step in the Na-abundance profile (we assume that [^{22}Ne/Na]$=0$). It is noted in [66] that ω Cen apparently does not strictly follow the global anticorrelation, since it contains giants with [Na/Fe]> 0.6. Therefore, the value $\delta M_{\rm mix} = 0.05$ was adopted when interpreting the observations for ω Cen in [66], and it

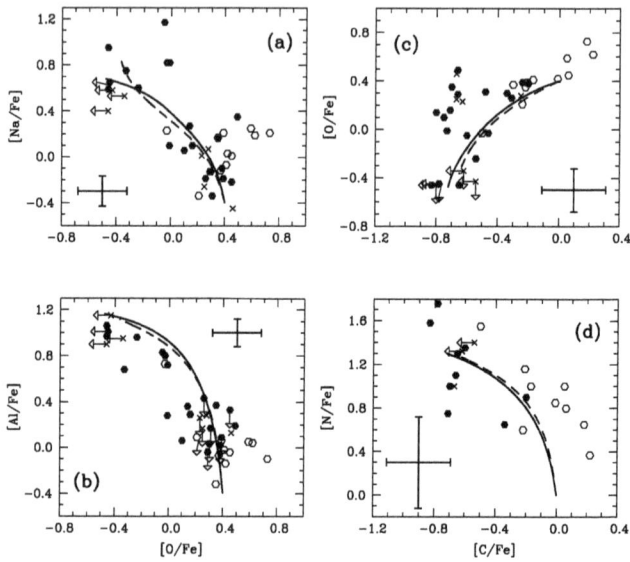

Figure 2.13. Correlation between abundances of C, N, O, Na and Al in the globular cluster ω Cen (descriptions of the symbols and notation, as well as the references, are given in Section 3.1) shown with the theoretical fits obtained using a semi-empirical diffusion model (dashed curve; the mixing depth and rate are $\delta M_{\mathrm{mix}} = 0.05$ and $D_{\mathrm{mix}} = 5 \cdot 10^8$ cm$^2 \cdot$s^{-1}) and a Zahn model, as in Fig. 2.12 (solid curves). The large crosses show the observational errors.

was concluded that in the case of ω Cen, we may be dealing with giants whose atmospheres are enriched in Na synthesised not only by ^{22}Ne but also partially by ^{20}Ne. On the other hand, the horizontal extent of the global anticorrelation reflects primarily the rate of the extra-mixing. To obtain O abundances as low as those observed in giants in M13, we must suppose that the mixing is very rapid (see the D_{mix} values used to interpret the observations for ω Cen and M13). In this sense, *globular clusters can be considered a unique "laboratory" for studies of various extra-mixing mechanisms.*

DEEP MIXING IN GLOBULAR-CLUSTER RED GIANTS

2.2.3 The Zahn Extra-Mixing Mechanism

Zahn [219] proposed an original description of how extra-mixing could arise and be supported in the radiative zone of a single, non-magnetic, rotating star. His only assumption was that the turbulence arising due to hydrodynamical instabilities associated with differential rotation of the star was highly anisotropic, with the horizontal component of the turbulent viscosity appreciably stronger than the vertical component. This main assumption brings about a transition to a state of shellular rotation in the star, in which the angular velocity Ω depends only on the distance from the centre r. In this case, the whole problem can be treated as being one-dimensional.

In a rotating star with the usual equation of state, the surfaces of constant entropy do not coincide with equipotential surfaces. Consequently, a spherically symmetrical flow of radiative energy cannot maintain a state of thermal equilibrium, leading to the appearance of (classical) meridional circulation [71, 206, 190]. The meridional circulation redistributes the angular momentum, and the initially uniform rotation becomes differential. Differential rotation can induce various hydrodynamical instabilities, of which Zahn [219] singled out the shear instability as having one of the shortest (dynamical) growth times. The thermal structure of the radiative zone hinders the growth of perturbations in the absence of differential rotation. If there is differential rotation, the situation must be analysed using the corresponding stability criteria (see the following section). If perturbations can grow, the resulting turbulent diffusion in the vertical direction will compete with the already present meridional circulation in the redistribution of the angular momentum. On the other hand, on the equipotential surfaces, nothing prevents perturbations from growing, and becoming sufficiently large to give rise to powerful horizontal turbulent diffusion, even in the presence of very weak shear flows. This last factor partially justifies Zahn's basic assumption of anisotropic turbulence.

We repeated the derivation of the equations of Maeder and Zahn [132] for the rate of meridional circulation in the presence of a molecular-weight gradient and non-stationary flows, since this situation is closest to the problem of extra-mixing in red giants. Taking into account minor corrections that do not appreciably influence the final numerical results and conclusions, we reproduce this derivation below.

The rate of meridional circulation

$$\vec{u} = u_r \vec{e}_r + u_\theta \vec{e}_\theta, \tag{2.7}$$

where $\vec{e}_r$ and $\vec{e}_\theta$ are unit vectors in a spherical co-ordinate system, appears in the equation of energy conservation, where it ensures local heat balance in the radiative zone of the rotating star:

$$\rho T \frac{\partial S}{\partial t} + \rho T (\vec{u}, \nabla) S = \nabla \cdot (\chi \nabla T) + \rho \varepsilon. \tag{2.8}$$

Here, S is the specific entropy, χ the coefficient of (radiative) thermal conductivity, and ε the rate at which nuclear energy is released per unit mass. We can obtain an approximation expression for $\vec{u}$ by solving a system of linearised equations made up of the equation of state, the energy equation, the Poisson equation, and the equation of hydrostatic equilibrium (taking into account the centrifugal force). Assuming that $\Omega = \Omega(r)$ and leaving only terms up to first order in the small parameter $(r\Omega^2)/g$ (g is the local gravitational acceleration) in the expansions of the structural variables of the stellar model in Legendre polynomials on the equipotential surfaces, we find, following Maeder and Zahn [132]

$$u_r(r,\theta) = U(r) P_2(\cos\theta), \quad u_\theta(r,\theta) = V(r) \frac{dP_2}{d\theta}, \tag{2.9}$$

where $P_2(\cos\theta) = \frac{1}{2}(3\cos^2\theta - 1)$ is a second-order Legendre polynomial. The amplitudes of the radial and latitudinal components of $\vec{u}$ are related by the continuity equation:

$$\frac{1}{r} \frac{d}{dr} [\rho r^2 U] - 6\rho V = 0. \tag{2.10}$$

The amplitude of the radial component, which is simply called the "rate of meridional circulation", is

$$U = U_0 \left[(E_\Omega + E_\mu) + \frac{C_P T}{\varepsilon_m \delta} \frac{\partial \Theta}{\partial t} \right], \tag{2.11}$$

where
$$U_0 = \frac{\varepsilon_m}{g}\left(\frac{P}{C_P \rho T}\right)\frac{1}{\nabla_{ad} - \nabla_{rad} + (\varphi/\delta)\nabla_\mu}. \qquad (2.12)$$

In formulae (2.11) and (2.12), $\varepsilon_m = L_r/M_r$ is the mean rate of production of total energy within a sphere with radius r and luminosity L_r, C_P is the specific heat capacity at constant pressure, and ∇_{ad} and ∇_{rad} are the adiabatic and radiative logarithmic temperature gradients. The molecular weight gradient is defined to be

$$\nabla_\mu = \frac{\partial \ln \mu}{\partial \ln P} = -\frac{H_P}{r}\frac{\partial \ln \mu}{\partial \ln r}, \qquad (2.13)$$

where $H_P = -\partial r/\partial \ln P$ is the pressure scale height.

The quantities

$$\delta = -\left(\frac{\partial \ln \rho}{\partial \ln T}\right)_{P,\mu}, \quad \varphi = \left(\frac{\partial \ln \rho}{\partial \ln \mu}\right)_{P,T} \qquad (2.14)$$

are uniquely defined by the equation of state. In particular, for a mixture of an ideal gas and radiation

$$P = P_{gas} + P_{rad} = \frac{\mathcal{R}}{\mu}\rho T + \frac{1}{3}aT^4 \qquad (2.15)$$

we have $\delta = (4-3\beta)/\beta$, where $\beta = P_{gas}/P$ and $\varphi = 1$. Zahn obtained the estimate

$$\Theta = \frac{\tilde{\rho}}{\rho} = \frac{2}{3}\frac{r}{g}\Omega^2 n, \quad n = \frac{\partial \ln \Omega}{\partial \ln r} \qquad (2.16)$$

for the quantity Θ, which measures the relative density fluctuations on an equipotential surface, $\tilde{\rho}/\rho$.

Maeder and Zahn proposed including the term $-\nabla \cdot \vec{F}_h$ on the right-hand side of the energy equation (2.8), where $\vec{F}_h \approx -D_h \rho T \nabla S$ is the heat flux transported by horizontal turbulence. In this case,

$$E_\Omega = E_\Omega^\star + 2\frac{H_T}{r}\frac{\rho_m}{\rho}\frac{D_h}{K}\left(\frac{\Theta}{\delta}\right), \qquad (2.17)$$

where $H_T = -\partial r/\partial \ln T$ is the temperature scale height, $\rho_m = M_r/(\frac{4\pi}{3}r^3)$ is the mean density, D_h is the coefficient of horizontal turbulent diffusion, and

$$K = \frac{\chi}{\rho C_P} = \frac{4acT^3}{3\kappa\rho^2 C_P} \tag{2.18}$$

is the coefficient of radiative energy diffusion.

Applying Zahn's transformation [219] further, following Maeder and Zahn [132], we find

$$\begin{aligned}
E_\Omega^\star &= 2\frac{\tilde{g}}{g}\left[1 - \frac{\Omega^2}{2\pi G\rho} - \frac{(\varepsilon + \varepsilon_g)}{\varepsilon_m}\right] \\
&- \frac{\rho_m}{\rho}\left\{\frac{r}{3}\frac{d}{dr}\left[H_T\frac{d}{dr}\left(\frac{\Theta}{\delta}\right) - \frac{\chi_T}{\delta}\Theta + (1 - \frac{1}{\delta})\Theta\right]\right. \\
&\left. - 2\frac{H_T}{r}\left(\frac{\Theta}{\delta}\right) + \frac{2}{3}\Theta\right\} \\
&- \frac{(\varepsilon + \varepsilon_g)}{\varepsilon_m}\left[H_T\frac{d}{dr}\left(\frac{\Theta}{\delta}\right) + (f_\varepsilon \varepsilon_T - \chi_T)\left(\frac{\Theta}{\delta}\right)\right. \\
&\left. + (2 - f_\varepsilon - \frac{1}{\delta})\Theta\right] + \frac{\varepsilon_g}{\varepsilon_m}\left(\Theta - \frac{\Theta}{\delta}\right), \tag{2.19}
\end{aligned}$$

and

$$\begin{aligned}
E_\mu &= \frac{\rho_m}{\rho}\left\{\frac{r}{3}\frac{d}{dr}\left[H_T\frac{d}{dr}\left(\frac{\varphi}{\delta}\Lambda\right) - (\chi_\mu + \frac{\varphi}{\delta}\chi_T + \frac{\varphi}{\delta})\Lambda\right]\right. \\
&\left. - 2\frac{H_T}{r}\frac{\varphi}{\delta}\Lambda\right\} + \frac{(\varepsilon + \varepsilon_g)}{\varepsilon_m}\left[H_T\frac{d}{dr}\left(\frac{\varphi}{\delta}\Lambda\right)\right. \\
&\left. + (f_\varepsilon \varepsilon_\mu + f_\varepsilon\frac{\varphi}{\delta}\varepsilon_T - \chi_\mu - \frac{\varphi}{\delta}\chi_T - \frac{\varphi}{\delta})\Lambda\right] \\
&+ \frac{\varepsilon_g}{\varepsilon_m}\frac{\varphi}{\delta}\Lambda. \tag{2.20}
\end{aligned}$$

In these equations,

$$\frac{\tilde{g}}{g} = \frac{4}{3}\frac{r\Omega^2}{g} \quad \Lambda = \frac{\tilde{\mu}}{\mu}, \tag{2.21}$$

where $\tilde{g}$ and $\tilde{\mu}$ (like $\tilde{\rho}$ earlier) are horizontal fluctuations of g and μ, ε_g is the rate of energy release (absorption) as a consequence of gravitational contraction (expansion), $f_\varepsilon = \varepsilon/(\varepsilon + \varepsilon_g)$ and

$$\chi_T = \left(\frac{\partial \ln \chi}{\partial \ln T}\right)_{P,\mu}, \quad \chi_\mu = \left(\frac{\partial \ln \chi}{\partial \ln \mu}\right)_{P,T}, \tag{2.22}$$

$$\varepsilon_T = \left(\frac{\partial \ln \varepsilon}{\partial \ln T}\right)_{P,\mu}, \quad \varepsilon_\mu = \left(\frac{\partial \ln \varepsilon}{\partial \ln \mu}\right)_{P,T}. \tag{2.23}$$

Unfortunately, the quantity D_h, which is currently important for the theory, cannot be determined unambiguously. Quite reasonably, we will assume that it is proportional to both U and V, since it is likely that the more rapid the circulatory flows, the stronger the differential motions and turbulence to which they give rise. Therefore, following Zahn [219], we will make use of the parametric expression

$$D_h = \frac{r}{C_h}\sqrt{(2V - \alpha U)^2 + U^2}, \tag{2.24}$$

where

$$\alpha = \frac{1}{2}\frac{d\ln(r^2\Omega)}{d\ln r} = 1 + \frac{1}{2}n, \tag{2.25}$$

and $C_h \leq 1$ is a free parameter.

In their numerical simulations, Chaboyer and Zahn [40] demonstrated that in the presence of strong horizontal turbulence, the mixing of elements by meridional circulation can be described purely as a diffusion process with the effective diffusion coefficient

$$D_{\text{eff}} = \frac{|rU|^2}{30 D_h}. \tag{2.26}$$

This result is valid only if $D_h \gg |rU|$, when the horizontal turbulence appreciably reduces the efficiency of the meridional circulation mixing compared to the classical case, when the diffusion coefficient is usually approximated as $|rU|$. Essentially, the horizontal turbulence "erodes" abundance inhomogeneities created on the equipotential surfaces via meridional circulation. After a time $t \gg r^2/(6D_h)$, this opposing action of the two mixing processes establishes a steady state in which

$$\Lambda = -\frac{rU}{6D_{\rm h}}\frac{\partial \ln\mu}{\partial \ln r} = \frac{rU}{6D_{\rm h}}\frac{r}{H_P}\nabla_\mu \tag{2.27}$$

(Maeder and Zahn [132]).

Meridional circulation is not the only process that redistributes angular momentum in the radius. As noted above, it competes with the vertical turbulent diffusion. In our model for extra-mixing in red giants, we computed the coefficient of vertical turbulent diffusion using the formula presented by Maeder and Zahn [131]:

$$D_{\rm v} = 2\frac{K}{N_T^2}(\frac{1}{5}\Omega^2 n^2 - N_\mu^2), \quad \text{if } \frac{1}{5}\Omega^2 n^2 > N_\mu^2, \tag{2.28}$$

where

$$N_T^2 = \frac{g\delta}{H_P}(\nabla_{\rm ad} - \nabla_{\rm rad}) \quad N_\mu^2 = \frac{g\varphi}{H_P}\nabla_\mu \tag{2.29}$$

are the squares of the T and μ components of the buoyancy frequency. The main assumption in the Zahn mechanism is essentially that $D_{\rm h} \gg D_{\rm v}$. We will discuss the problem of making an appropriate choice of expression for $D_{\rm v}$ in the following section.

Since meridional circulation and (vertical) turbulent diffusion act jointly, the right-hand side of the equation of angular momentum transport includes two terms:

$$\frac{\partial}{\partial t}(\rho r^4 \Omega) = \frac{1}{5}\frac{\partial}{\partial r}\left[\rho r^4 \Omega(U - 5\dot{r})\right] + \frac{\partial}{\partial r}\left(\rho r^4 \nu_{\rm v}\frac{\partial \Omega}{\partial r}\right). \tag{2.30}$$

The first term takes into account meridional circulation and the second the diffusion of angular momentum together with vertical turbulence [132]. We assume the coefficient of turbulent viscosity $\nu_{\rm v}$ in (2.30) to be equal to $D_{\rm v}$. Equation (2.30) is written in Euler coordinates, so that its flow term contains the local rate of mass flow $\dot{r} = (\partial r/\partial t)_{M_r}$. The contributions of meridional circulation and vertical turbulence in mixing of elements is adequately described by (2.6) if we set

$$D_{\rm mix} = D_{\rm eff} + D_{\rm v} \tag{2.31}$$

in this equation.

The Zahn mechanism yields a self-consistent solution for the problem of rotation-induced mixing in the radiative zone of a star, in the sense that the rotation profile is no longer taken to be arbitrary (for example, $\Omega(r)$ = const, as has often been assumed in earlier studies). Instead, the evolution of the rotation profile brought about by the redistribution of angular momentum in the star is computed.

2.2.4 Choice of Criterion for Shear Instability

As we will show in the following section, the radiative zone between the helium core and the base of the convective envelope in a red giant is probably in a state of strongly differential rotation. Under these conditions, when the horizontal component of the flow velocity $\mathcal{U}$ varies with depth $z = R - r$, where R is the stellar radius, Kelvin–Helmholtz (shear) instability can arise (Chandrasekhar [41]). It is possible to obtain a general expression for the criterion for shear instability by considering the motion of two gas elements separated by a small vertical distance δz. The horizontal velocities of the elements differ by $(d\mathcal{U}/dz)\delta z$. In order for the elements to change places, the force of inertia must work against the force of gravity. If the work required is less than the difference in the kinetic energies of the gas elements in their initial and final positions, such an exchange of positions will be energetically allowed, and consequently the corresponding shear flow is unstable. The following criterion for shear instability can serve as a quantitative expression for these arguments:

$$\frac{1}{4}\left(\frac{d\mathcal{U}}{dz}\right)^2 > \frac{g}{\rho}\left(\frac{d\rho'}{dz} - \frac{d\rho}{dz}\right), \qquad (2.32)$$

where $\rho'(z)$ describes the variation of the density inside a gas element during its motion. Using the equation of state $\rho = \rho(P, T, \mu)$, we can rewrite (2.32)

$$\frac{1}{4}\left(\frac{d\mathcal{U}}{dz}\right)^2 > \frac{g\delta}{H_P}\left[(\nabla' - \nabla) + \frac{\varphi}{\delta}(\nabla_\mu - \nabla'_\mu)\right], \qquad (2.33)$$

where ∇ and ∇_μ are the gradients of T and μ in the surrounding medium and ∇' and ∇'_μ are the same gradients inside the gas element.

The criterion (2.33) remains unsuitable for practical use until the quantities ∇, ∇' and ∇'_μ are determined. The simplest assumptions in this connection would be (i) $\nabla = \nabla_{\text{rad}}$, meaning that the motion brought about by the shear instability does not influence the thermal structure of the surrounding medium; and (ii) $\nabla' = \nabla_{\text{ad}}$ and iii) $\nabla'_\mu = 0$, i.e. there is no exchange of heat or matter between the gas elements and the surrounding medium.[4] Under these assumptions, the condition (2.33) is transformed into the conservative form

$$\frac{1}{4}\left(\frac{d\mathcal{U}}{dz}\right)^2 > \frac{g\delta}{H_P}\left[\nabla_{\text{ad}} - \nabla_{\text{rad}} + \frac{\varphi}{\delta}\nabla_\mu\right]. \tag{2.34}$$

The dashed curve in Fig. 2.14 depicts the base-ten logarithm of the ratio of the right-hand and left-hand sides of (2.34) as a function of the mass δM for the computation results discussed in the following section. We can see that when condition (2.34) is used, the entire radiative zone of the red giant is stable against the growth of perturbations produced by shear instability.

However, the gas element can be considered to be moving adiabatically only if its lifespan (the mixing time) is much shorter than the characteristic time for radiative losses from its surface (an element deep inside the star can be taken to be optically thick). The inverse ratio of these two timescales is known as the Peclé number $Pe = lv/K$, where l and v are the characteristic size and velocity of the element. Thus, the assumption $\nabla' = \nabla_{\text{ad}}$ is admissible only when $Pe \gg 1$. In mixing length theory, this limit corresponds to the case of very efficient convective heat transport – adiabatic convection [51].

Mixing length theory [129] gives

$$\nabla' - \nabla = \frac{\Gamma}{\Gamma+1}(\nabla_{\text{ad}} - \nabla), \tag{2.35}$$

where $\Gamma = Pe/6$ measures the efficiency of vertical heat transport by an element of gas. During the heat transport process, the gas element perturbs the thermal structure of the surrounding medium.

---

[4] In fact, as will be shown below, assumptions (i) and (ii) cannot be considered independent.

DEEP MIXING IN GLOBULAR-CLUSTER RED GIANTS 67

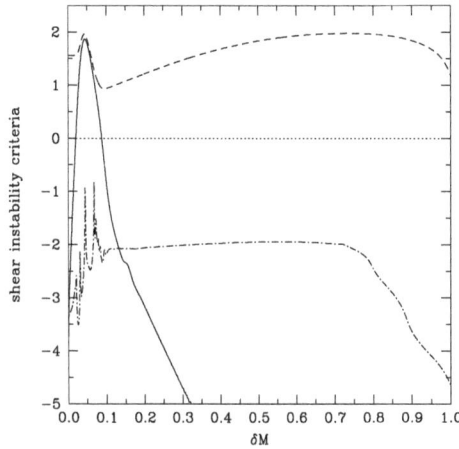

Figure 2.14. Base-ten logarithm of the ratio of the right-hand and left-hand sides of equations (2.34) (dashed) (2.38) (solid) and (2.42) (dot–dashed), representing various criteria for shear instability.

This perturbation can be taken into account via an appropriate choice of ∇. Using the same mixing length theory, Maeder [129] obtained

$$\nabla = \frac{\nabla_{\rm rad} + \dfrac{6\Gamma^2}{\Gamma + 1}\nabla_{\rm ad}}{1 + \dfrac{6\Gamma^2}{\Gamma + 1}}. \qquad (2.36)$$

Together with the instability criterion (2.33), relations (2.35) and (2.36) form a closed system, which can be used to estimate the coefficient of turbulent diffusion as follows [131]. We first substitute equation (2.35) into (2.33), assuming in the latter that $\nabla'_\mu = 0$. We can now find $\Gamma_{\rm max}$, such that the inequality (2.33) is satisfied when $\Gamma < \Gamma_{\rm max}$. In other words, the differential flow under consideration is marginally stable when $\Gamma = \Gamma_{\rm max}$. We now make another approximation, replacing the entire turbulence spectrum contributing to D_v with the single mode $(lv)_{\max} = 6K\Gamma$; i.e. we assume $D_v \approx \frac{1}{3}(lv)_{\max} = 2K\Gamma$. Further, following Maeder and Meynet [131], we assume that the main contributions to the vertical heat transport and to D_v are made by the same turbulent elements. This makes it possible to exclude the unknown ∇ from the expression for $\Gamma_{\max}$ using (2.36). As a result, we obtain a quadratic equation for $\Gamma_{\max}$, whose solution depends on the parameter

$$\theta = \frac{\frac{\varphi}{\delta}\nabla_\mu - \frac{1}{4}\frac{H_P}{g\delta}\left(\frac{d\mathcal{U}}{dz}\right)^2}{\nabla_{\mathrm{ad}} - \nabla_{\mathrm{rad}}}. \tag{2.37}$$

If $\theta \geq 0$, then, according to Maeder and Meynet [131], shear instability will not arise. Thus, when we take into account radiative losses and vertical heat transport by the moving gas element, the criterion for shear instability takes the form

$$\theta < 0, \quad \frac{1}{4}\left(\frac{d\mathcal{U}}{dz}\right)^2 > \frac{g\varphi}{H_P}\nabla_\mu. \tag{2.38}$$

The logarithm of the ratio of the right-hand and left-hand sides of this inequality is illustrated in Fig. 2.14 by the solid curve. We can see that the appearance of shear instability, and consequently of vertical turbulence and diffusion, are now possible in the region between the base of the convective envelope and layers with $\delta M \approx 0.10$.

The results of the computations described in the following section indicate that everywhere above the point $\delta M = 0.10$, the value of the parameter $-\theta$ is of the order 10^{-2}. In the limit $-\theta \ll 1$, Maeder and Meynet [131] give the estimate $\Gamma_{\max} \approx -\theta$ and

$$D_v \approx 2K\frac{\frac{1}{4}\left(\frac{d\mathcal{U}}{dz}\right)^2 - \frac{g\varphi}{H_P}\nabla_\mu}{\frac{g\delta}{H_P}(\nabla_{\mathrm{ad}} - \nabla_{\mathrm{rad}})}. \tag{2.39}$$

This same formula was used in our application of the Zahn mechanism to describe extra mixing in red giants. Indeed, our expression (2.28) can be obtained from (2.39) directly after substituting

$(d\mathcal{U}/dz)^2 = (r\sin\theta d\Omega/dr)^2$ and averaging over the equipotential surface:

$$\overline{\left(\frac{d\mathcal{U}}{dz}\right)^2} = \frac{\int_0^\pi \left(\frac{d\mathcal{U}}{dz}\right)^2 \sin^3\theta\, d\theta}{\int_0^\pi \sin^3\theta\, d\theta} = \frac{4}{5}\left(r\frac{d\Omega}{dr}\right)^2. \qquad (2.40)$$

Maeder and Meynet applied the criterion (2.38) in their preliminary investigation of rotation-induced mixing in massive MS stars. They established that according to this criterion, mixing could not penetrate deep enough to transport sufficient quantities of helium to the stellar surface to explain the observations of Herrero et al. [91].

As we can see, a similar "negative" result is obtained in the case of globular-cluster red giants. Indeed, $\delta M \approx 0.10$ is greater than the value $\delta M_{\mathrm{mix}} \approx 0.08$ required for extra-mixing to reach layers in which the O abundance begins to decrease (Fig. 2.11). As in massive MS stars, the main factor preventing deeper penetration of the turbulent diffusion in red giants is the barrier produced by the gradient of μ. At the same time, we cannot avoid noting that the resulting depth $\delta M \approx 0.10$ already guarantees that the extra-mixing will reach layers in which both CN-cycle reactions and the synthesis of Na from ^{22}Ne occur.

In their attempts to "ease" the passage of turbulent diffusion through the μ-gradient barrier, Talon and Zahn [195] and Maeder [130] introduced certain additional assumptions about the physics of the interactions between the moving gas elements and the surrounding medium. In both cases, they tried to somehow change the internal chemical composition of the gas element; i.e. they essentially speculated on the possibility that $\nabla'_\mu \neq 0$.

Using an analogy with heat exchange between the surrounding medium and an element during its motion as a consequence of radiation, Talon and Zahn [195] suggested that horizontal turbulence can bring about a similar exchange of mass. In this case, the molecular weight of the rising element would decrease while that of the sinking element would increase, relative to their initial values. As a consequence, the inertial forces would be required to perform less work against the gravitational force. Talon and Zahn obtained the following expression for the turbulent diffusion coefficient:

$$D_v = \frac{\frac{1}{2}\left(\frac{d\mathcal{U}}{dz}\right)^2}{N_T^2/(K+D_h) + N_\mu^2/D_h}, \tag{2.41}$$

which should be used under the condition that

$$\left(\frac{d\mathcal{U}}{dz}\right)^2 > \frac{2}{3}\nu Re_c \left[\frac{N_T^2}{K+D_h} + \frac{N_\mu^2}{D_h}\right], \tag{2.42}$$

where ν is the usual (molecular plus radiative) viscosity and $Re_c \approx 3000$ is the critical Reynolds number. The logarithm of the ratio of the right- and left-hand sides of (2.42) is shown in Fig. 2.14 by the dot–dashed curve. We can see that, if this criterion were applicable in red giants, extra-mixing should be present throughout the radiative zone, contrary to observations. In particular, such "overly deep" mixing would absorb the second step in the Na-abundance profile (Fig. 2.11), leading to atmospheric excesses of Na in red giants larger than those observed. In addition, carrying so much helium outward, such mixing would obviously fundamentally change the entire course of evolution on the red-giant branch [212]. Talon–Zahn extra-mixing also appears improbable from the physical point of view: it is difficult to imagine that rising and sinking gas elements could survive the penetration of vortices of horizontal turbulence. They are probably completely disrupted by such vortices in the course of their motion. A more correct analogy with heat loss by a moving gas element via radiation would be a microscopic diffusion of matter, but this occurs too slowly to be of importance.

Maeder [130] proposed the alternative hypothesis that additional sources of turbulence in the star (horizontal turbulence brought about by shear instability, or semi-convection in massive MS stars) could give rise to local fluctuations in μ that make a gas element lighter (heavier) than the surrounding medium at the beginning of its motion. This could stimulate spontaneous rising (sinking) of the element. Maeder used a different method than that outlined above to derive a formula for the diffusion coefficient applicable in this situation. He first modified the formula of Kippenhahn and Weigert

([102], p. 45) for the so-called sinking velocity v_μ, taking into account the gradient of μ (v_μ is the velocity of a gas element whose internal molecular weight has suddenly become higher than μ in the surrounding medium, so that it has begun to sink). The diffusion coefficient was estimated as $D = \frac{1}{3}|v_\mu|d$, where d is the diameter of the gas element. However, in our view, Maeder's approach is not internally self-consistent. For example, when deriving his final formula (5.30) for D, he used relations (3.16) and (4.26), which were obtained for the cases of two mutually-exclusive assumptions about thermodynamic variations in the moving element. The first of these was obtained assuming that $D\rho = \rho' - \rho \neq 0$, while the second was based on the assumption that $D\rho = 0$. In particular, this internal inconsistency means that Maeder's new criterion for shear instability

$$\frac{1}{4}\left(\frac{dU}{dz}\right)^2 > \frac{g\delta}{H_P}(\nabla' - \nabla) \qquad (2.43)$$

reduces to the old criterion (2.38) if we take into account the relation $\nabla' - \nabla = (\varphi/\delta)\nabla_\mu$, which follows directly from the equilibrium condition $DT/T = (\varphi/\delta)D\mu/\mu$ used by Maeder when deriving his expression for v_μ.

It is likely that only three-dimensional hydrodynamic computations can provide a conclusive answer to the rather complex question of the development of shear instability in the radiative zones of stars in the presence of a gradient in μ.

2.2.5 Results of Numerical Computations

The evolution of a star with initial mass $M = 0.8\,M_\odot$, initial helium content $Y = 0.24$, heavy-element abundance $Z = 5 \cdot 10^{-4}$ and ratio of the mixing length to the pressure scale height $\alpha = l_{\rm conv}/H_P = 1.48$ was computed from the zero age main sequence to the onset of the core helium flash using the program of Raffelt and Weiss [151]. Three red giant models were selected from the evolutionary sequence, as described in Section 2.2.2. The main parameters of these models related to the question of extra-mixing are presented in Table 2.1. We used these models to interpolate T, ρ, r and $D_{\rm mix}$ in δM and $\lg L/L_\odot$ in the solution of the system of nuclear reaction kinetics

Table 2.1. Parameters of three basis red-giant models

$\log_{10} L$	R	r_{bce}	r_c	Ω_{bce}	v_{bce}	v^j	v^Ω
2.08	15.6	1.13	0.0250	8.0	6.3	0.46	87
2.70	37.4	1.35	0.0245	5.6	5.3	0.19	147
3.34	95.2	1.96	0.0234	2.7	3.7	0.08	180

Luminosities and all radii are given in solar units, Ω_{bce} is in units of 10^{-6} rad·s^{-1} and the linear rotational velocities are in km·s^{-1}. The superscripts "j" and "Ω" indicate that the corresponding specific angular momentum and angular rotational velocity were taken to be constant throughout the convective envelope.

equations (2.6). The code for solving Eqs. (2.6) is described in [66]. We first determine the diffusion coefficient D_{mix} as a function of δM for each of the three selected basis models using the following algorithm.

(1) The hydrogen abundance distribution in the radiative zone between the helium core and the base of the convective envelope $X(\delta M)$ is calculated assuming the absence of extra-mixing.

(2) The calculations show that the gas pressure dominates over the radiative pressure throughout the radiative zone, in accordance with the assumption that $\delta = 1$ in all formulas.

(3) Due to the non-linear relationship between Λ and U (it follows from (2.24) that D_h in (2.27) depends on both U and V; in accordance with (2.10), V is determined by U and dU/dr, while U itself is a complex function of Λ via E_μ (Eqs. (2.11) and (2.20))) Λ is found in two steps. In the first, we assume that $D_h \approx |rU|$, so that $\Lambda \approx \pm(1/6)(r/H_P)\nabla_\mu$ [132]. This provides a first approximation for E_μ, U and V and a second approximation for D_h (with $C_h = 0.5$ in (2.24)). This second approximation is much higher than the first approximation $D_h \approx |rU|$, especially near the hydrogen burning shell, where U changes sign several times within a very narrow region (Fig. 2.16c), so that V takes on very large values. This substantially decreases Λ in the second step. Using the new value for Λ (Fig. 2.16a), we again calculate E_μ, U (Fig. 2.16b and 2.16c), V and D_h.

(4) We can see in Table 2.1 that during the evolution of a red giant, the radii of its helium core r_c and the base of its convective envelope r_{bce} change only slightly compared to the corresponding variations in the surface radius R. Therefore, we can use Eq. (2.30) in Euler co-ordinates as an approximation. The boundaries of the radiative zone retain approximately the same radii, and there is an inward flow of matter in this zone that feeds the hydrogen burning shell. The rate of this flow appears in Eq. (2.30), and can be estimated as

$$\dot{r} = -\frac{1}{4\pi r^2 \rho}\frac{L}{EX_{env}}, \qquad (2.44)$$

where X_{env} is the hydrogen abundance in the convective envelope and $E \approx 6.3 \cdot 10^{18}$ erg·g^{-1} is the energy released by hydrogen burning.

(5) It is desirable to relate Ω_{bce} to the surface rate of rotation of the star. We consider two simple laws for the angular velocity variations in the convective envelope: $\Omega \propto r^{-2}$, valid for constant specific angular momentum $j = \frac{2}{3}r^2\Omega =$ const; and rigid rotation, $\Omega =$ const. The corresponding linear rotational velocities at the surface are given in Table 2.1 as v^j and v^Ω. The latter are much higher than the observed values, and even exceed the escape velocities for the two models corresponding to the largest scales. Kumar, Narayan and Loeb [110] pointed out that the appearance of the stationary rotation profile in the convective zone of a star depends on the interaction between convective elements. Solid body rotation is supported only in the presence of purely elastic isotropic scattering of convective elements by each other, while fully inelastic scattering leads to the distribution $j =$ const. We are again forced to conclude that a final solution to the problem of choosing the correct rotation profile in the convective envelope of a red giant can be obtained only by three-dimensional calculations. For the time being, Ω_{bce} remains a free parameter whose value cannot be unambiguously related to the surface rotational velocity. The only thing we can expect is a positive correlation between Ω_{bce} and the surface rotation.

(6) It turns out that the correlations between the abundances of various elements in giants in ω Cen can be theoretically reproduced fairly well if the angular rotational velocity at the base of the convective envelope in the initial model is taken to be $\Omega_{bce} = 8 \cdot 10^{-6}$ rad·s^{-1}. In the remaining two models, Ω_{bce} was determined by the condition that $j_{bce} = \frac{2}{3}r_{bce}^2 \Omega_{bce}$ be constant.

(7) Equation (2.30) is solved using the Henyey method with the boundary conditions

$$n = \frac{\partial \ln \Omega}{\partial \ln r} = -2, \quad U = 0 \quad r = r_c$$

and

$$n = -2, \quad \Omega = \Omega_{bce} \quad r = r_{bce}.$$

We choose the initial rotation profile with the form $\Omega = \Omega_{bce}(r_{bce}/r)^2$. Note that if we ignore all terms on the right-hand side of Eq. (2.30) except for the flow term with $\dot{r}$, we obtain as a stationary solution satisfying these boundary conditions j = const, as expected. However, it is important to answer the question of whether or not the joint action of meridional circulation and turbulent diffusion appreciably changes the uniform distribution of the specific angular momentum that is obtained in their absence. Our answer is "No"!

(8) We trace the evolution of $\Omega(r)$ in the radiative zone of a red giant by solving (2.30) for each of the three basis models. The solution time intervals for the first and second models (12.7 and 4.6 million years) were taken to be half the differences between the ages of the first and second, and the second and third models, respectively. The corresponding time interval for the third model, which is very close to the tip of the red giant branch, was taken to be $5 \cdot 10^5$ years. As indicated above, we found that meridional circulation and turbulent diffusion only slightly perturbed the initial profile $\Omega \propto r^{-2}$ on these time intervals. Figure 2.15 shows the final rotation profiles. There are only appreciable variations in the Ω profiles in the second and third models near the hydrogen burning shell, where they are due to

DEEP MIXING IN GLOBULAR-CLUSTER RED GIANTS   75

variations in U in a narrow region adjacent to the helium core. These U variations are due to a growth in the μ gradient and the ratio ρ_m/ρ with depth, as well as the variable contribution of ε_g to $E_\Omega^\star$ and E_μ in (2.19) and (2.20). We emphasise that ε_g plays an important role in red giants due to the presence of mass flows.

(9) Finally, using (2.26) and (2.28), we compute the coefficient D_{mix} (2.31) as a function of the relative mass. The resulting distributions $D_{\mathrm{mix}}(\delta M)$ for the three selected models are then used to interpolate the values D_{mix} in our calculations of nucleosynthesis with extra-mixing.

Figure 2.16d illustrates the quantity $\lg D_{\mathrm{mix}}$ as a function of δM for the first model (in the second and third models, D_{mix} is larger by approximately one and two orders of magnitude respectively). At the base of the convective envelope, down to $\delta M(D_v = 0) \approx 0.10$, where D_v becomes equal to zero, the behaviour of D_{mix} is determined primarily by D_v; i.e. the contribution of meridional circulation to the total diffusion coefficient is negligible. It is important to note the following two circumstances in this connection. First, it turns out that in the Zahn mechanism, the main origin of extra-mixing in the bulk of the red giant's radiative zone is (vertical) turbulent diffusion, not meridional circulation, as suggested by Sweigart and Mengel [192] and Smith and Tout [177]. Second, the horizontal turbulent diffusion appreciably lowers the efficiency of mixing by meridional circulation, via erosion of the chemical composition. The coefficient D_h dominates all other diffusion coefficients. If there were no such horizontal erosion, D_{eff} would be larger than D_v.

However, our criterion for shear instability (2.38) does not allow vertical turbulent diffusion to penetrate deeper than $\delta M(D_v = 0) \approx 0.10$. Meridional circulation is responsible for any extra-mixing that occurs below this point. However, the circulation rate U changes sign when $\delta M(U = 0) = 0.068$. It is striking that this depth falls in the interval $0.06 < \delta M_{\mathrm{mix}} < 0.07$ indicated by the morphology of the global anticorrelation of [O/Fe] with [Na/Fe] (see Section 2.2.2). Therefore, we adopt as a working hypothesis the idea that the depth of extra-mixing in a red giant is determined by the position of the first point

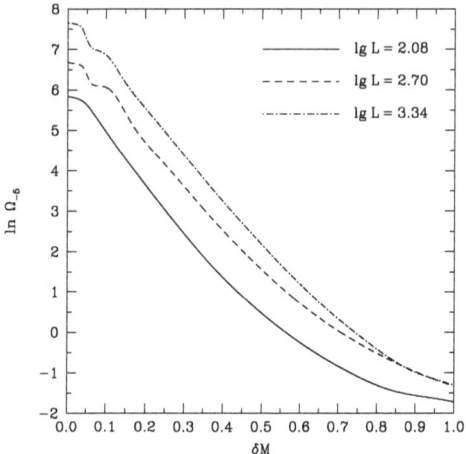

Figure 2.15. Final profiles of the angular rotational velocity (arbitrarily shifted along the vertical axis, $\Omega_{-5} = \Omega/10^{-5}$ rad·s^{-1}) in the three consecutive basis red giant models (the corresponding luminosity is given in the top right-hand corner) selected for the interpolation of the distributions of mixing and structural parameters of stars (see text). These profiles were obtained from the initial profile $\Omega \propto r^{-2}$ by solving Eq. (2.30). The integration times for the first ($12.7 \cdot 10^6$ years, solid curve) and second ($4.6 \cdot 10^6$ years, dashed curve) models were taken to be half the differences between the ages of the consecutive models. The integration time for the third model, located near the tip of the red giant branch, was $5 \cdot 10^5$ years (dot–dashed curve).

below $\delta M(D_{\rm v} = 0)$ where U changes sign, i.e. $\delta M_{\rm mix} = \delta M(U = 0)$. These depths for the three basis models are $\delta M_{\rm mix} = 0.068, 0.066$ and 0.075. Of course, we should bear in mind that various sources of errors in the calculations, especially those associated with calculating $D_{\rm h}$ and U (note the lack of monotonic behaviour in U in Fig. 2.16c), could slightly change these values for $\delta M_{\rm mix}$. Nevertheless, we believe that the closeness of these values to the semi-empirical estimates of $\delta M_{\rm mix}$ is unlikely to be a coincidence. It is more probable that this reflects the reasonable nature of our assumptions about the dependence

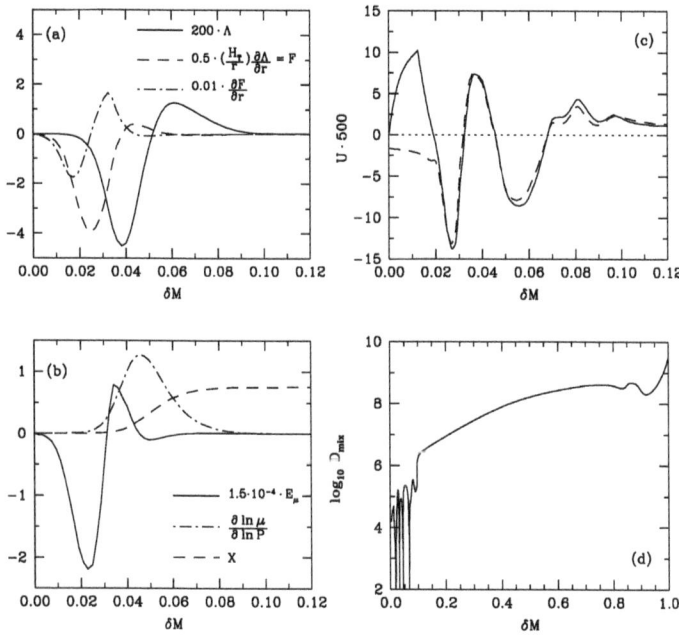

Figure 2.16. Some results of our numerical calculations of extra-mixing in red giants with the Zahn mechanism for the initial model ($\lg L/L_\odot = 2.08$): (**a**) scaled parameter Λ (formula 2.27) and its derivative; (**b**) scaled quantity E_μ (formula 2.20), μ gradient and hydrogen content; (**c**) scaled rate of meridional circulation (U in cm·s^{-1}) for the initial (dashed curve) and final (solid curve) profiles Ω (the first point on the right where $U = 0$ determines the mixing depth); (**d**) total diffusion coefficient. An appreciable contribution to D_{mix} is made by vertical turbulent diffusion below the base of the convective envelope, down to layers with $\delta M \approx 0.10$, where D_v becomes equal to zero. Between $\delta M(U = 0) = 0.068$ and $\delta M(D_\mathrm{v} = 0) \approx 0.10$, the only acting mechanism for extra-mixing remains meridional circulation.

of $\delta M_{\rm mix}$ on physical quantities such as the gradient of μ and the profiles of $\varepsilon_{\rm g}/\varepsilon$ and $\rho_{\rm m}/\rho$ near the hydrogen burning shell. The points $\delta M(D_{\rm v}=0)$ and $\delta M(U=0)$ for the first model are marked by the vertical line segments along the horizontal axis in Fig. 2.11.

The validity of the assumptions underlying the first step of our algorithm was confirmed *a posteriori*. We compared the life spans of protons in the CNO cycle $\tau_{\rm p}({\rm CNO})$ with the characteristic timescale for extra-mixing $\tau_{\rm mix} \sim r^2/D_{\rm mix}$ near the hydrogen burning shell. In the second and third models, $\tau_{\rm p}({\rm CNO})/\tau_{\rm mix} < 1$ everywhere below $\delta M(D_{\rm v} = 0)$. The physical reason for this is that the strong horizontal turbulence inside the hydrogen burning shell makes the transport of material by circulation (the only mixing process that can act below $\delta M(D_{\rm v} = 0)$) very inefficient. This justifies our use of models without mixing to determine $\delta M(U = 0)$. At the same time, it is possible that a deeper part of the hydrogen burning shell could be chemically uniform due to mixing by meridional circulation in stars near the tip of the red giant branch. The very high values of $D_{\rm h}$ in the hydrogen burning shell also justifiy use of the stationary solution for Λ [step (3)].

The results of our calculations of nucleosynthesis with diffusion coefficient $D_{\rm mix}$ obtained using the above algorithm are shown by the solid curves in Figs. 2.12 and 2.13. We can see in Fig. 2.13 that the Zahn mechanism and semi-empirical diffusion model provide a good agreement with the observations for ω Cen. In Fig. 2.12, the theoretical curve for the Zahn mechanism reproduces the global anticorrelation between [O/Fe] and [Na/Fe] very well for all globular clusters except M13. The mixing must be about five times more rapid in the case of M13 (the diffusion coefficients required for the semi-empirical model to explain the observational data for ω Cen and M13 are $D_{\rm mix} = 5 \cdot 10^8$ and $D_{\rm mix} = 2.5 \cdot 10^9$ cm$^2 \cdot$s^{-1}, respectively). Such an increase in $D_{\rm mix}$ demands an increase in $\Omega_{\rm bce}$ by only a factor of $\sqrt{5} \approx 2.2$, since $D_{\rm v} \propto \Omega^2$ (2.28). However, we do not claim at this stage to be able to find precise values of $\Omega_{\rm bce}$ corresponding to the observations. For example, formula (2.28) gives only a lower limit for $D_{\rm v}$, since we neglected the contribution of turbulent vortices with $lv < (lv)_{\rm max}$ in its derivation. If the turbulence spectrum obeys a Kolmogorov law, then lv varies in proportion to $l^{4/3}$, and the contribution of vortices with $l < l_{\rm max}$ could be important.

2.2.6 Conclusion

The Zahn mechanism can reproduce the observed correlations between elemental abundances in globular clusters fairly well. In this mechanism, extra-mixing in the radiative zone of a rotating red giant results from the joint action of meridional circulation and turbulent diffusion. The horizontal turbulence, which is thought to be on a much larger scale than the vertical turbulence, supports a state of shellular rotation $\Omega = \Omega(r)$ via the smoothing of any differential flows on equipotential surfaces. It is also responsible for a horizontal erosion of chemical composition, making it possible to describe mixing of material by meridional circulation as a purely diffusion process. Vertical turbulent diffusion competes with meridional circulation in the radial redistribution of angular momentum. If there were neither meridional circulation nor turbulent diffusion, a uniform distribution of the specific angular momentum would rapidly be established in the red giant's radiative zone, as a consequence of the quasi-stationary inward flow of mass. Due to the large difference in the radii of the inner and outer boundaries of the radiative zone $r_{\rm bce}/r_{\rm c} > 45$ (Table 2.1), this would imply the presence of strongly differential rotation with $\Omega \propto r^{-2}$. Our calculations show that meridional circulation and turbulent diffusion are not able to appreciably change this steep profile of Ω over the characteristic evolution time for a red giant (Fig. 2.15). Note, however, that the action of the dynamo effect may be able to disrupt such differential rotation [200], even in the presence of a weak magnetic field. Unfortunately, verification of this hypothesis must await the results of the corresponding magneto-hydrodynamical calculations.

In the bulk of the radiative zone, the total diffusion coefficient $D_{\rm mix}$ is determined primarily by the coefficient of vertical turbulent diffusion $D_{\rm v}$. Below the point δM where $D_{\rm v}$ becomes equal to zero, the only mixing mechanism in action is meridional circulation, and here $D_{\rm mix} \approx D_{\rm eff} = |rU|^2/(30 D_{\rm h})$. In our application of the Zahn mechanism to red giants, there is essentially only one free parameter – the angular rotational velocity at the base of the convective envelope $\Omega_{\rm bce}$ ($C_{\rm h}$ should be of order unity [219], and, as shown by our calculations, variations in this parameter leave the main results virtually unaffected). $\Omega_{\rm bce}$ is determined by the mixing rate, since $D_{\rm mix} \approx D_{\rm v} \propto \Omega^2 \propto \Omega_{\rm bce}^2$. We associated the mixing depth with the position of the first point (going inward) below the depth $\delta M(D_{\rm v} = 0)$ where the meridional circulation changes sign, $\delta M_{\rm mix} = \delta M(U = 0) < \delta M(D_{\rm v} = 0)$. It is important to note that to the first approximation, this value of $\delta M_{\rm mix}$ does not depend on $\Omega_{\rm bce}$. Instead, it is sensitive to the properties of the red giant structure (the distributions of T, ρ, μ, ε and $\varepsilon_{\rm g}$). The lack of dependence of $\delta M_{\rm mix}$ on $\Omega_{\rm bce}$ can be understood if we compare the E_Ω and E_μ terms in (2.11). Near the hydrogen burning shell, where $\delta M_{\rm mix}$ is usually located, E_μ dominates over E_Ω, and E_μ is completely independent of Ω. The Zahn mechanism yields values for $\delta M_{\rm mix}$ that are surprisingly close to our previous semi-empirical estimates obtained using a simple diffusion model in which the physical mechanism for the extra-mixing was not specified.

We found that the correlations between elemental abundances in giants in ω Cen and the global anticorrelation between [O/Fe] and [Na/Fe] in M3, M15 and M92 were best produced when $\Omega_{\rm bce} = 8 \cdot 10^{-6}$ rad·s^{-1} in the initial red giant model. Rotation with roughly twice this speed is required for the giants in M13.

In principle, $\Omega_{\rm bce}$ can be related to the linear rotational velocity at the surface, v. The last two columns of Table 2.1 give values of v for the cases of rigid rotation (v^Ω) and constant specific angular momentum (v^j) in the convective envelope. Rigid rotation of the envelope with the velocity $\Omega_{\rm bce}$ for which the calculations were carried out leads to a centrifugal acceleration at the surface that exceeds the gravitational acceleration. If we suppose that the convective envelope rotates differentially according to the law $j = $ const, the resulting velocity $v = v^j$ is more realistic for red giants. An alternative solution is the absence of any definite relationship between the slowly rotating envelope and rapidly rotating core. In the absence of a well-established dependence between the rotation at the surface and $\Omega_{\rm bce}$, we are forced to treat this last quantity as a free parameter, though we expect there to be some positive correlation between v and $\Omega_{\rm bce}$.

As noted in Section 2.2.1, the most rapidly rotating blue horizontal branch stars were found in M13 [145]. Among these, six stars have $v \sin i \geq 30$ km·s^{-1}. Blue horizontal branch stars with line-of-sight rotational velocities from 2 to 20 km·s^{-1} have been found in M3 [145]. This difference in the $v \sin i$ values for M13 and M3 is in qualitative

DEEP MIXING IN GLOBULAR-CLUSTER RED GIANTS 81

agreement with our conclusion that the red giants in M13 should rotate approximately twice as fast as the giants in M3, M15 and M92. If we adopt $T_{\text{eff}} \approx 10^4$ K and $L \approx 10^2 \, L_\odot$ as estimates of the effective temperature and luminosity of blue horizontal branch stars, we obtain the radius estimate $R \approx 3 \, R_\odot$. Applying the law $j = \text{const}$ to the phase of transition of the star from the red giant branch to the horizontal branch after the core helium flash, we can translate v^j in Table 2.1 into $v_{\text{HB}} \approx 3$ km·s^{-1}. This crude estimate is close to the observed rotational velocities of blue horizontal branch stars, justifying our choice of $\Omega_{\text{bce}} = 8 \cdot 10^{-6}$ rad·s^{-1} for the initial red giant model satisfying the correlations between elemental abundances of giants in ω Cen, M3, M15 and M92.

There are a number of pieces of evidence that the same extra-mixing mechanism (the Zahn mechanism) acts in the radiative envelopes of massive MS stars (see Section 4.1). Talon et al. [196] and Denissenkov et al. [67] showed that, after a relatively short time (compared to the main-sequence life span of the star), a massive MS star with sufficiently rapid initial rotation makes a transition to a stationary regime in which meridional circulation and vertical turbulent diffusion compensate each other almost exactly. In this regime, the mixing of material takes place primarily via turbulent diffusion (as in red giants). The characteristic mixing time, especially near the boundary of the convective core, proves to be much longer than the main-sequence life span in all cases considered [67]. This leads to a delay in the appearance of anomalies in the abundances of CNO elements at the surface of a massive MS star (time is required for the "wave of diffusion" to reach the atmosphere), which may explain the delay in the increase of the surface He abundance observed in OB stars [126]. Note that although we used formula (2.41) to calculate D_v in [67], due to the additional assumptions $K \gg D_h$ and $N_\mu^2 = 0$ made in [67], this formula is essentially identical to (2.28).

2.3 Episodic Production of Lithium in Red Giants (2000)

2.3.1 Introduction

Lithium (here and below we refer only to the isotope ^{7}Li) is an extremely "fragile" element: it is destroyed at temperatures as low as $T \sim 2.5 \cdot 10^6$ K. Therefore, in MS stars with masses $M \geq 1 \, M_\odot$,

Li can be preserved only in the outer 1–4% of the total mass of the star [161], in the absence of mixing that could transport it to hotter, deeper layers. It is thought that the convective envelope of a MS star with mass $M < 0.9\,M_\odot$ is sufficiently deep for a significant quantity of Li to burn in it by the end of the star's life span on the main sequence. It is also supposed that the Li content in a star already decreases appreciably during the phase of evolution preceding the main sequence, with the decrease being greater the lower the mass of the star and the higher its metal content (elements heavier than helium) [147].

However, these theoretical predictions are in contradiction with the measured Li abundances of Population I dwarfs in two respects. First, the observations indicate the existence of a so-called "dip" in the Li abundances of type F dwarfs with effective temperatures $6400 \leq T_{\rm eff}({\rm K}) \leq 6900$, with this "lithium dip" being deeper in older stars. The origin of the "hot" side of this dip was recently explained successfully by Talon and Charbonnel [194] as a consequence of extra-mixing due to meridional circulation and turbulent diffusion supported by the stellar wind. They attributed the increase in the Li abundance on the "cool" side of the lithium dip to the action of another (as yet unknown) mechanism for the mixing of stellar material and transport of angular momentum, which apparently becomes more efficient in cool dwarfs. Second, the observed Li deficits in old Population I dwarfs with effective temperatures $T_{\rm eff} \leq 6000$ K are much higher than the values predicted by standard stellar models, and there is also a large scatter in the Li abundances [147]. In particular, the solar lithium abundance is roughly two orders of magnitude lower than the abundance in meteorites, in which $\log\varepsilon({\rm Li}) \equiv \log[n({\rm H})/n({\rm Li})] + 12.0 = 3.3$ [2]. This value is usually taken to be the characteristic primordial Li abundance in Population I stars. This solar Li-abundance anomaly may be a consequence of the same sort of extra-mixing acting on the "cool" side of the lithium dip, and may also bring about the nearly rigid rotation of the Sun's core indicated by helio-seismological measurements [194].

The Li abundance anomaly is less pronounced in metal-poor dwarfs ([Fe/H] ≤ -1). For example, in place of a dip, a fairly distinct "Li plateau" is observed, extending in effective temperature from 5600 K to 6400 K [186]. The mean Li content in the region of the plateau is approximately $\log \varepsilon(\mathrm{Li}) = 2.3$, which is considered characteristic of both the primordial Li abundance of Population II stars and the chemical composition of the primordial matter formed in the Big Bang [197, 13, 87]. In cooler Population II dwarfs, the observed lithium deficit is higher than expected, as in Population I dwarfs.

Fortunately, the aforementioned uncertainties in the evolution of the Li content of stars on the main sequence have almost no influence on the problem discussed in this chapter, since the observations show that the physical processes behind it lead to the destruction and not the creation of lithium. Summarising this introduction, the atmospheric Li abundances in stars leaving the main sequence should not exceed their primordial values, and, at best, the amount of Li that can be present is only that initially contained in the outer 1–4% of the total mass of the star.

2.3.2 Extra-Mixing in Red Giants

After its time on the main sequence, a low-mass star ($M \leq 2.5\,M_\odot$) first moves to the subgiant branch, then begins its first ascent along the red giant branch. In these evolutionary phases, the star consists of a degenerate helium core surrounded by a hydrogen burning shell and an extended convective envelope. The hydrogen burning shell and the base of the convective envelope are separated by a thin (in mass, but not in radius) zone that is in a state of radiative equilibrium. The base of the convective envelope begins to move inward in the star on the sub-giant branch. This marks the onset of the so-called "first dredge-up", when the surface chemical composition of the star is subject to modest variations, which depend on the initial mass of the star and its metallicity. These variations are due to convective mixing of material in the atmosphere with inner layers of the star, where nuclear reactions of the pp chains and CNO cycle occurred on the main sequence. Indicators of the first dredge-up include, for example, the ratio $^{12}\mathrm{C}/^{13}\mathrm{C}$ and the Li content; both should decrease during the evolution of the star in this phase. We also expect a modest decrease in the C abundance and an increase in the N abundance [14]. It is important to note that the surface abundance of ^{3}He grows substantially (by roughly an order of magnitude) during the first dredge-up. It is known that this isotope is produced in large quantities in the periphery of the MS star, where non-equilibrium pp-chain reactions lead to the formation of a "bump" in the distribution of the ^{3}He abundance. On the red giant branch, convection "smears" this bump over the whole envelope. Lithium is simply diluted by a factor of 20–60 in the envelope of the red giant [161] as a consequence of convective mixing of the outer layers, where the Li content either remains constant on the main sequence and retains its primordial value, or decreases slightly in inner layers, where Li was already burning on the main sequence.

The first dredge-up ends when the base of the convective envelope ceases its inward motion in the star and begins to retreat outward. According to standard stellar evolution theory, the surface chemical composition should subsequently no longer change on the red giant branch. This theoretical prediction is not supported by the observations. In low-mass field red giants, the surface abundances of C, N and Li, and also the ratio ^{12}C/^{13}C continue to vary right up to the end of the star's evolution on the giant branch. Moreover, the abundances of O, Na, Mg and Al in globular-cluster red giants are subject to similar variations (see Sections 2.1 and 2.2). These last three elements play the role of catalysts in reactions of the NeNa and MgAl cycles, which operate in parallel with the CNO cycle that is responsible for the energy release in the hydrogen burning shell. This discrepancy between the standard theory and observations has been successfully explained in red giant models with extra-mixing in the radiative zone between the hydrogen burning shell and the base of the convective envelope [192, 61, 42, 43, 64, 66, 63]. It is usually believed that extra-mixing can act efficiently in the radiative zone only if the chemical composition in this zone is uniform. Therefore, it is usually assumed that extra-mixing in red giants can begin only after the outward-moving hydrogen burning shell has smoothed the jump in hydrogen abundance left earlier by the base of the convective envelope at the time of its deepest penetration into the stellar interior. Following Gratton *et al.* [87], we will refer to the part of the red giant branch containing stars in which the hydrogen burning shell has already crossed this abundance jump as the "upper red giant branch". The base of the upper red giant branch is roughly at $\log L/L_\odot = 2$. Standard evolutionary calculations show that the hydrogen burning shell reaches the hydrogen-abundance jump while the star is still on the red giant branch only in low-mass stars. This is in excellent agreement with the fact that observational manifestations of extra-mixing are displayed only by low-mass red giants [42].

Recently, Gratton et al. [87] determined the abundances of Li, C, N, O and Na and the ratio $^{12}C/^{13}C$ for a large sample of field stars with well-established luminosities and metal contents in the interval $-2 \leq$ [Fe/H] ≤ -1. Fig. 2.17 compares the values of $\log\varepsilon(^{7}Li)$, [C/Fe], $\log^{12}C/^{13}C$ (here, as above, the chemical symbol for an isotope denotes its density) and [N/Fe] for these stars (squares) with the results of our calculations obtained using the method and code described in [64], in which we modelled extra-mixing using diffusion (solid curve). We can see that the behaviour of the observed quantities on the upper red giant branch (at $\log L/L_\odot > 2$) is well fitted by the model with diffusion mixing reaching a depth $\delta m_{\mathrm{mix}} = 0.12$ and having a rate (diffusion coefficient) $D_{\mathrm{mix}} = 5 \cdot 10^8$ cm$^2 \cdot$s^{-1} (δm or δM is a relative mass co-ordinate, such that $\delta m = 0$ at the base of the hydrogen burning shell where the hydrogen content is $X = 10^{-4}$ and $\delta m = 1$ at the base of the convective envelope). The dotted and dashed lines in Fig. 2.17 encompass the range of luminosities corresponding to the MS phase and the first dredge-up, respectively.

Based on the upper part of Fig. 2.17, we can conclude that *(i)* the initial Li content is preserved in the atmospheres of a large fraction of Population II MS stars, *(ii)* there is a dilution of Li during the first dredge-up, to abundances close to those predicted by the standard theory and *(iii)* there is convincing evidence for extra-mixing on the upper red giant branch, which reduces the surface abundance of Li in stars with $\log L/L_\odot > 2$.

In contrast to the situation for Population II field stars, in which neither deficits of O nor excesses of Na are observed, star-to-star variations in the abundances of both O and Na are observed in globular clusters. More importantly, in globular-cluster red giants, [Na/Fe] increases as [O/Fe] decreases (Fig. 2.18, various symbols). The existence of this global anticorrelation between [O/Fe] and [Na/Fe] has also been explained as a result of extra-mixing (see [63] and references therein), however, deeper mixing is required in this case than for field giants (the solid curve in Fig. 2.18 was calculated for $\delta m_{\mathrm{mix}} = 0.06$, $D_{\mathrm{mix}} = 5 \cdot 10^8$ cm$^2 \cdot$s^{-1}). The evolution of the surface Li abundance

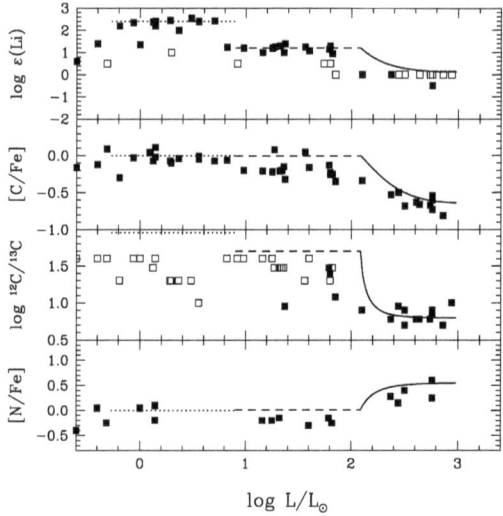

Figure 2.17. Li, C and N abundances and the ratio $^{12}C/^{13}C$ for field stars with well established luminosities and metal contents in the interval $-2 \leq [\text{Fe/H}] \leq -1$ [87]. The white squares show upper (for Li) or lower (for $^{12}C/^{13}C$) observational limits. The dotted and dashed horizontal line segments show the ranges of luminosities for MS stars and for the first dredge-up, respectively. The solid curves were obtained from calculations modelling extra-mixing with diffusion reaching a depth $\delta m_{\text{mix}} = 0.12$ and with a rate $D_{\text{mix}} = 5 \cdot 10^8$ cm$^2 \cdot$s^{-1}.

corresponding to the solid curve in Fig. 2.18 is depicted by curve 3a in Fig. 2.20. Note that in all three cases (Fig. 2.17, solid curve; Figs. 2.18 and 2.20), the calculations of the nuclear reaction kinetics with extra-mixing started with the same red-giant model, with $M = 0.8\,M_\odot$, $\log L/L_\odot = 2.1$ and heavy element content $Z = 5 \cdot 10^{-4}$ ($[\text{Fe/H}] \approx \log Z/Z_\odot = -1.58$). The corresponding profiles for the ^{3}He, ^{7}Be, ^{7}Li, ^{12}C, ^{13}C and Na abundances in the initial model are shown in Fig. 2.19 (curve without jumps at $\delta m \approx 0.14$). Because the

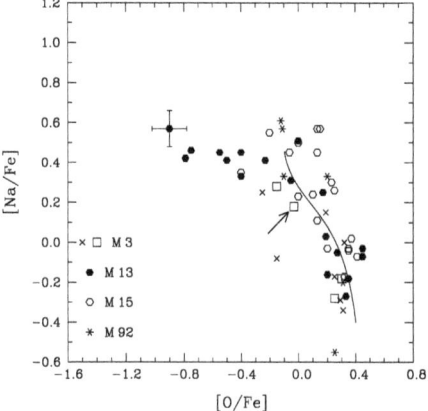

Figure 2.18. Global anticorrelation of [O/Fe] with [Na/Fe] for globular-cluster red giants and theoretical fit for a diffusion model for extra-mixing with depth $\delta m_{\mathrm{mix}} = 0.06$ and rate $D_{\mathrm{mix}} = 5 \cdot 10^8$ cm$^2 \cdot$s^{-1} (solid line). The white squares show data for giants of M3 from the recent study of Li abundances by Kraft et al. [104]. The unusually lithium-rich giant IV-101 is marked with an arrow.

extra-mixing in the models for field giants is thought to be less deep ($\delta m_{\mathrm{mix}} = 0.12$), the surface Na abundance in these models does not change, in accordance with observations [87].

2.3.3 The Problem of Lithium-Rich Red Giants

In the vast majority of field giants (K giants), the surface abundances of Li are, indeed, very low, in agreement with the predicted dilution of Li during the first dredge-up and its subsequent destruction in association with extra-mixing [23]. However, there is a small group of K giants (a few percent of the population) that have remarkably high Li abundances, sometimes even exceeding the primordial value [209, 89, 23, 11, 54, 55, 56]. Only recently has the evolutionary status of lithium-rich giants begun to become clear. Based on

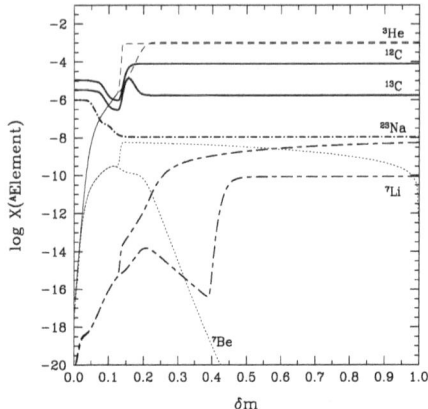

Figure 2.19. Abundance profiles for some nuclides in the radiative zone of the starting model (see text). The abundance variations for ^{3}He, ^{7}Be and ^{7}Li over $6.3 \cdot 10^4$ yrs produced by extra-mixing with depth $\delta m_{\mathrm{mix}} = 0.14$ and rate $D_{\mathrm{mix}} = 5 \cdot 10^{11}$ cm$^2 \cdot$s^{-1} are shown (profiles with jumps at $\delta m \approx 0.14$).

accurate luminosity estimates obtained via parallax measurements using the Hipparcos satellite, Jasniewicz et al. [97] concluded that most lithium-rich giants have passed through the first dredge-up (as confirmed by their low ^{12}C/^{13}C ratios). Consequently, the primordial Li abundance in their atmospheres could not have been preserved.

Generally speaking, any scenario explaining the existence of these "lithium giants" must somehow incorporate one of the following basic ideas: (1) preservation of a high initial Li abundance in some way; (2) transport of Li to the envelope of the giant by some external source; (3) production of "fresh" Li within the red giant itself. De la Reza et al. [56], Jasniewicz et al. [97] and others prefer the internal production of Li. Since their arguments are rather convincing, we will not discuss them here. In both these studies, the most probable mechanism for the internal production of Li is the so-called Cameron–Fowler ("^{7}Be-transport") mechanism [28, 29]. In this case, it is supposed that nuclei of ^{7}Be synthesised in the reaction $^3\text{He}(\alpha,\gamma)^7\text{Be}$ are rapidly transported to outer, cooler layers of the star, enabling them to avoid destruction in the reaction chains $^7\text{Be}(p,\gamma)^8\text{B}(e^+\nu)^8\text{Be} \to 2\alpha$ and $^7\text{Be}(e^-,\nu)^7\text{Li}(p,\alpha)^4\text{He}$. In these cooler layers, only the reaction $^7\text{Be}(e^-,\nu)^7\text{Li}$ can take place, completing the Li-production cycle in the Cameron–Fowler mechanism (see the curves with jumps at $\delta m \approx 0.14$ in Fig. 2.19). It is important to note that scenarios based on idea (3) above have recently obtained new, weighty support: Castilho et al. [35] discovered that Be was strongly underabundant (by more than 90%) in several lithium giants they investigated.

Following their discovery that the Cameron–Fowler mechanism could operate in a natural way in bright, intermediate-mass stars on the AGB [160], Sackmann and Boothroyd [161] recently showed that, under certain conditions, this mechanism could also produce Li in stars on the red-giant branch. Freshly synthesised ^{7}Be in AGB stars is carried outward by ordinary convection, while the presence of extra-mixing is required in red giants. As will be shown below, the main problem that arises in attempts to explain the origin of lithium giants using the Cameron–Fowler mechanism is that the properties of the extra-mixing required are very unusual.

It is interesting that most field lithium giants probably have circumstellar dust envelopes [55]. Moreover, a large number of new lithium giants were discovered using an IRAS two-colour diagram, among stars with infrared excesses. In the opinion of De la Reza et al. [55], this property is probably the only feature distinguishing lithium giants from ordinary K giants. They even proposed a scenario relating the Li excesses in the studied stars with evolution of the circumstellar envelope. In this scenario, *every* low-mass red giant passes through a short phase in its evolution when an unidentified internal mechanism enriches its atmosphere in lithium and simultaneously gives rise to intense flows of material from the stellar surface. Using a simple model, De la Reza et al. [55] calculated evolutionary tracks (on an IRAS two-colour diagram) for an envelope that separates from the star and is scattered in the interstellar medium. They concluded that the total duration of the proposed phase is about 10^5 yrs; a very rapid initial stage (lasting only a few thousand years) with a sudden increase in the surface Li abundance accompanied by

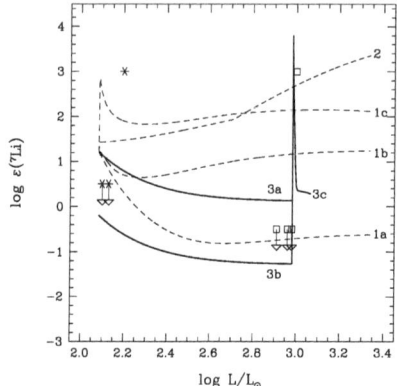

Figure 2.20. Evolution of the surface Li abundance in stars on the upper red giant branch due to extra-mixing. The following mixing parameters were used in the calculations: **1**: $\Delta \log T = \log T(\delta m = 0) - \log T(\delta m_{\mathrm{mix}}) = 0.26$ ($\delta m_{\mathrm{mix}} \geq 0.14$), **a** – $D_{\mathrm{mix}} = 10^9$ cm$^2 \cdot$s^{-1}; **b** – $D_{\mathrm{mix}} = 5 \cdot 10^9$ cm$^2 \cdot$s^{-1}; **c** – $D_{\mathrm{mix}} = 10^{11}$ cm$^2 \cdot$s^{-1}; **2**: $\Delta \log T = 0.36$ ($\delta m_{\mathrm{mix}} \geq 0.17$), $D_{\mathrm{mix}} = 10^{11}$ cm$^2 \cdot$s^{-1}; **3 a,b**: $\delta m_{\mathrm{mix}} = 0.06$, $D_{\mathrm{mix}} = 5 \cdot 10^8$ cm$^2 \cdot$s^{-1}; **3c**: δm_{mix} decreasing by 0.01 every $8 \cdot 10^3$ years from 0.16 to 0.06, after which a constant value of 0.06 is maintained, with the mixing rate always constant and equal to $D_{\mathrm{mix}} = 10^{12}$ cm$^2 \cdot$s^{-1}. The squares show the Li abundance for giants in the globular cluster M3 according to the data of Kraft et al. [104], and the asterisks show the data of Hill and Pasquini [92] for the metal-poor open cluster Berkeley 21.

the ejection of the envelope is followed by a much more prolonged period (to 10^5 yrs), in which the envelope is scattered in the interstellar medium and, more importantly, the Li abundance in the atmosphere of the giant simultaneously decreases.

2.3.4 *Proposed Solution to the Lithium Problem*

Recently, Siess and Livio [172] proposed another original scenario based on idea (2) above: the red giant absorbs a brown dwarf or giant planet (for brevity, simply "planet") orbiting around it, in which the primordial Li abundance should have been preserved, since the mass of such bodies is too small for the onset of thermonuclear reactions (the only exception may be deuterium burning). The absorption of this body brings its preserved Li into the giant's envelope, and also leads to the ejection of a small fraction of the envelope as a consequence of thermodynamic processes associated with the merger (accretion of material near the base of the convective envelope, where the planet is expected to "disintegrate", could lead to the thermal expansion of above-lying layers, in turn accelerating the star's mass loss rate; for more detail see [172]). Unfortunately, this scenario has an obvious shortcoming – it cannot explain Li abundances exceeding the characteristic primordial values observed in a number of giants.

In [65], Denissenkov and Weiss proposed a combined scenario based on ideas (2) and (3): it is possible that the absorption of a planet by the red giant triggers the Cameron–Fowler mechanism within the star. Our scenario is based on new observational data showing for the first time unusually high Li abundances in giants belonging to very metal-poor clusters. Such giants include IV-101 ([Fe/H] = -1.50) in the globular cluster M3 [104] and T33 ([Fe/H] = -0.58) in the open cluster Berkeley 21 [92]. In both cases, the lithium abundances are $\log \varepsilon(^7\text{Li}) \approx 3.0$. The lithium giants IV-101 and T33 are shown on the evolutionary diagram in Fig. 2.20 together with five normal giants from the same clusters (the squares show the data of Kraft et al. [104] for M3 and the asterisks the data of Hill and Pasquini [92] for Berkeley 21). We can see from the observations that an episodic enrichment of the stellar atmosphere in lithium can occur at any point on the upper red giant branch. Indeed, the total evolution of the star along the upper red giant branch lasts approximately $3 \cdot 10^7$ yrs, much longer than the $\sim 10^5$ yrs required for the completion of the entire cycle for the production and subsequent destruction of Li proposed by De la Reza et al. [55]. Figure 2.20 shows that in T33, the Cameron–Fowler mechanism was initiated near the base of the upper red giant branch, while it was initiated near the tip of this branch in the case of IV-101. This independence of the onset of lithium enrichment in the stellar atmosphere from the star's evolutionary stage provides evidence that it is an external source that turns on this mechanism. Figure 2.18 supports this conclusion: two giants in M3 have similar [Na/Fe] and [O/Fe] values, in full agreement with the general run of the global anticorrelation, but one of them (marked with an arrow) is rich and the other poor in lithium. At the same time, extra-mixing with parameters chosen to reproduce the global anticorrelation (Fig. 2.18, solid curve) is not able to enrich the envelope of a giant in lithium (Fig. 2.20, curves 3a and 3b). All the evidence suggests that after a fairly prolonged period ($\sim 3 \cdot 10^7$ yrs) of "ordinary" extra-mixing typical of globular-cluster red giants, with depth $\delta m_{\mathrm{mix}} \approx 0.06$ and rate $D_{\mathrm{mix}} \approx 5 \cdot 10^8$ cm$^2 \cdot$s^{-1}, some event occurred in IV-101, which suddenly varied the extra-mixing parameters such that they became favourable for the production of lithium.

In test calculations using stellar models corresponding to the lithium giants IV-101 and T33, we established that the Cameron–Fowler mechanism is capable of efficiently synthesising Li and then maintaining the Li abundance at a high level over a fairly prolonged time only if the depth of the extra-mixing lies in the range $0.12 \leq \delta m_{\mathrm{mix}} \leq 0.18$ and, more importantly, if the mixing is very rapid, $D_{\mathrm{mix}} \geq 10^{11}$ cm$^2 \cdot$s^{-1} (Fig. 2.20, curves 1c and 2). Recently, Denissenkov and Tout [63] showed that if a number of conditions are satisfied, the Zahn mechanism involving meridional circulation and turbulent diffusion brought about by rotation can serve as the physical mechanism for extra-mixing in globular-cluster red giants [219, 132]. However, our calculations indicate that mixing rates $D_{\mathrm{mix}} \geq 10^{11}cm^2 \cdot$s^{-1} can be obtained in the Zahn mechanism only as upper limits, and only if the layers of the star rotate with velocities that are close to the local Keplerian rates.[5] Such rapid rotation can be understood in a natural way in a scenario with absorption of a planet – as a result of the transfer of the orbital angular momentum of the absorbed body to the envelope of the red giant (for more detail, see [172]). The next question that must be answered is how to obtain the correct values of the extra-mixing depth in such a scenario.

The dashed curves in Fig. 2.20 are analogous to those depicted in Fig. 10 in [161]. They were calculated for constant values of the depth and rate of extra-mixing, which are favourable for the production of Li. In fact, one of these (similar to our curve 2) was used by Smith

---

[5] According to N.A. Drake (private communication), on average, lithium giants rotate more rapidly than normal K giants.

DEEP MIXING IN GLOBULAR-CLUSTER RED GIANTS 93

et al. [180] to interpret the data for a lithium giant located near the tip of the upper red giant branch of the globular cluster NGC 362. However, such a simple interpretation cannot be fully correct, since *(i)* it does not provide the enrichment of the giant's envelope in Na or depletion in O observed for IV-101; *(ii)* it can explain neither the decrease in the Li abundance that follows its production nor the fairly short duration of the overall cycle observed for lithium giants [55]; *(iii)* it requires a very unusual combination of mixing parameters, corresponding to shallow but very rapid mixing compared to the values that can reproduce the global anticorrelation betweeen [O/Fe] and [Na/Fe] (we believe it would be more natural to expect that more rapid mixing should simultaneously be deeper).

We suggest the following explanation for how a scenario with absorption of a planet can give rise to the correct mixing depth. According to Siess and Livio [172], the planet is disrupted near the base of the red giant's convective envelope. Further, the rotation profile in the radiative zone acquires a step-like form, with a jump increase in the angular rotational velocity to a value close to the local Keplerian rate in the layer where the final destruction of the planet occurred. In the subsequent evolution of the star, the hydrogen burning shell continues to move outward, and after approximately $8 \cdot 10^5$ yrs (the time taken for the shell to cross the entire radiative zone) it reaches the jump in the rotation profile. Not long before this, over $\sim 8 \cdot 10^4$ yrs, the jump in the rotation profile will cross the region with relative depths $0.06 \leq \delta m \leq 0.16$, where (and when) the production of Li becomes efficient. In this way, we do not fix the mixing depth to be in the range favourable for the production of Li. Instead, we propose that very rapid extra-mixing (induced by the absorption of a planet) produces Li when the lower boundary of the zone in which it acts passes through the region of favourable values of δm (Fig. 2.20, curve 3c). The advantage of this scenario is that it can explain both the production of Li and the subsequent decrease in its abundance. Indeed, calculations show that, after the depth of the rapid mixing (jump in the rotation profile) becomes less than $\delta m_{\mathrm{mix}} \sim 0.08$, the Li abundance in the envelope of the giant begins to decrease over a timescale of $\sim 10^5$ yrs, in agreement with the results of [55]. Note that it is even more difficult to provide a sufficiently rapid decrease in the surface Li abundance after its jump-like production than it is to produce the Li in the first place, but this occurs in a natural way in our proposed scenario.

2.3.5 Conclusion

We must note, however, that there exist lithium giants whose origin cannot be explained in our scenario. One example is V42 in the globular cluster M5, with a lithium abundance $\log \varepsilon(^7\text{Li}) \approx 1.8$ [33], which is apparently a low-mass star in a phase in its evolution immediately after the AGB. The timescales for the cycle of lithium enrichment and its subsequent depletion in the giant's atmosphere ($\sim 10^5$ yrs) and for the evolution on the red-giant branch ($\sim 10^7$ yrs) rule out the possibility that the high Li abundance in V42 remains from the time when the star was still a red giant. On the other hand, since the luminosity of V42 ($M_{\text{Bol}} = -3.38$) does not exceed the luminosities of stars near the tip of the red giant branch, but it is much hotter, and consequently smaller, than such red giants, it is natural to suppose that its absorption of a planet must have occured when it was a red giant, not in the recent past ($\leq 10^5$ yrs). Thus, our scenario cannot work for V42, although it is otherwise a typical representative of M5; in particular, its atmosphere shows a standard excess of α elements (^{16}O, ^{24}Mg, etc.), as well as the usual anticorrelation between [O/Fe] and [Na/Fe] [33]. The only suggestion we can make in relation to this star is that the enrichment of its atmosphere in lithium occured via the action of the Cameron–Fowler mechanism, but when the star was on the AGB, where the extra-mixing gave rise to "hot bottom burning" at the base of the convective envelope. This view is now considered standard, and is thought to lead to the synthesis of Li in intermediate-mass AGB stars [160]. Due to the very thin ($< 10^{-2} M_\odot$) envelope of V42, even modest extra-mixing could be sufficient in this case.

Our scenario also has no direct connection with the formation of a circumstellar dust envelope. Siess and Livio [172] relate the ejection of the envelope with an increase in the stellar mass-loss rate during its absorption of a planet, but, in our scenario, this event is separated in time (by about $7 \cdot 10^5$ yrs) from the onset of the Li production. It seems to us that the solution of this problem should be sought in one of the following ways.

(1) The mass of the radiative zone in the red giant is negligibly small compared to the mass of the helium core, while the angular rotational velocity in this zone varies as r^{-2} (see Denissenkov and Tout [63] and Section 2.2 of this book). Consequently, the ratio of the centrifugal acceleration to the gravitational acceleration grows with depth as r^{-1}. Since we expect this ratio to be close to unity near the base of the convective envelope immediately after absorption of the planet, in the subsequent inward motion of the jump in the rotation profile, this ratio must exceed unity somewhere deep in the radiative zone, possibly leading to dynamic processes carrying the excess angular momentum outward. This could accelerate the rate of mass loss, but with some time delay.

(2) The absorption of a planet itself could be accompanied by various dynamic and thermodynamic processes, such as an increase in the depth of the convective envelope [172], which could redistribute material containing higher angular momentum throughout the radiative zone. In this case, there should not be a time delay, and the rapid extra-mixing can reach depths favourable for the production of Li from its very onset.

The question of whether or not these proposed alternative scenarios are realised in actual red giants can be addressed theoretically only via three-dimensional hydrodynamic calculations.

3. Chemical Evolution of Globular Clusters

3.1 A Combined Scenario: Inherited Chemical Anomalies Plus Extra-Mixing in Red Giants (1997)

3.1.1 Introduction

In 1996, Shetrone [170] determined the isotopic composition of magnesium for a small sample of bright red giants in the globular cluster M13. The fact that a large fraction of the giants investigated by him has $[(^{25}\text{Mg}+^{26}\text{Mg})/^{24}\text{Mg}] \approx +0.4$, and not -0.4, like most halo stars (see [133]), is a result of fundamental importance, and confirms our general understanding of differences between the chemical compositions of globular-cluster and field stars. The work of Shetrone followed the earlier discovery that, in M13, larger excesses of Al are usually accompanied by modest underabundances of Mg [169]. It was also demonstrated earlier that the anticorrelation of Na and Al with O is present not only in "normal mono-metallic" clusters, like M13 [107], but also in the massive globular cluster ω Cen, whose stars display an initial scatter in metallicity and in the abundances of s-process elements (see the 1995 work of Norris and Da Costa [139], henceforth NDC95). These new data present a challenge for stellar evolution theory while providing additional observational evidence about possible primordial nucleosynthesis processes that could be responsible for star-to-star variations in surface elemental abundances in globular clusters. In this section, these data are used to address the question of whether the modern theory of nucleosynthesis in stars is able to reproduce the entire spectrum of variations of surface abundances in globular-cluster red giants in a self-consistent way, and, if it is not, what additional assumptions about the structure and evolution of globular-cluster stars are required.

Stars leaving the main sequence in modern globular clusters have modest masses, $M \approx 0.8 - 0.9\,M_\odot$, and low metallicities in the range $-2.4 \leq [\text{Fe/H}] \leq -0.2$, which corresponds to $8 \cdot 10^{-5} < Z < 0.01$, if we adopt $[\text{Fe/H}] = \lg(Z/Z_\odot)$, where $Z_\odot = 0.01886$, in accordance with [2]. In the standard theory of stellar evolution, there are no doubts about the subsequent structural and chemical evolution of such stars, at least up to the onset of the core helium flash. The burning of hydrogen at the centre on the main sequence, during which pp-chain reactions dominate, is now relegated to a hydrogen burning shell, where the main source of energy is the CNO cycle. As it moves outward, this shell brings about a gradual growth of the mass of the helium core it surrounds. This continues until the core is massive enough to ignite a helium flash. These qualitative changes in the internal structure of the star are accompanied outwardly by its rise along the red giant branch on the Hertzsprung–Russell (HR) diagram. The surface chemical composition of the star should not change substantially in this evolutionary phase. The only important event in this connection between the star's departure from the main sequence and the core helium flash is the well-known first dredge-up, which begins on the sub-giant branch. The motion of the star from the main sequence to a cooler region in the HR diagram favours the development of

DEEP MIXING IN GLOBULAR-CLUSTER RED GIANTS   97

convection in the stellar envelope. The base of the convective envelope penetrates deeper and deeper into the star, finally reaching layers that were in a state of radiative equilibrium on the main sequence, and in which modest changes in the abundances of isotopes of C and N occurred on the main sequence, in reactions of the then energetically insignificant CN cycle. As a result, the surface abundances of ^{12}C, ^{13}C and ^{14}N begin to deviate from their initial values (the abundances of ^{7}Li, ^{3}He and a modest number of other light nuclides not discussed here are also affected; see [161, 214, 43] for results of calculations of variations of their abundances during the first dredge-up and in the presence of deep extra-mixing). The motion of the base of the convective envelope into the stellar interior continues to some maximum depth, marking the end of the first dredge-up, after which the convection begins to retreat, and the base of the envelope gradually shifts outward in proportion to the analogous motion of the hydrogen burning shell. Our standard calculations for the evolution of a star with $M = 0.8\,M_\odot$ show that, depending on Z, the variations of the surface abundances of ^{12}C, ^{13}C and ^{14}N in the first dredge-up are very modest: $(Z, ^{12}\text{C}/^{13}\text{C}, \Delta\lg^{12}\text{C}, \Delta\lg^{14}\text{N}) = (10^{-4}, 64, -0.0024, 0.0025)$, $(5 \cdot 10^{-4}, 50, -0.0064, 0.012)$, $(5 \cdot 10^{-3}, 45, -0.0092, 0.021)$ (on the MS, $^{12}\text{C}/^{13}\text{C} \approx 90$).

Comparisons of these predictions of the standard theory of stellar evolution with available observational data on the atmospheric chemical compositions of globular-cluster red giants reveal numerous evident disagreements. *(i)* Measurements of the ratio $^{12}\text{C}/^{13}\text{C}$ are very low, often being close to the limit of 3.5 reached by the CN cycle in equilibrium [176, 21, 189, 24, 18, 170]. *(ii)* Differences in the C and N abundances in red giants within a single cluster reach an order of magnitude and more. Moreover, [C/Fe] is anticorrelated with luminosity (more precisely, it is correlated with M_V), in agreement with a gradual decrease in the C abundance as a star rises along the red giant branch after the end of the first dredge-up (similar anticorrelations have been found, for example, for stars in M92 and NGC 6397 [9, 117, 17], as well as in M4 and NGC 6752 [189]). *(iii)* The most intriguing result is that there are rather large (reaching 1 dex) star-to-star variations in the O and Na abundances in many globular clusters (see the review by Kraft [103]); such variations are also observed for Al and Mg in some clusters [107].

It is important that all those nuclides whose abundances show appreciable scatter from star to star in globular clusters (^{12}C, ^{13}C, N, O, Na, isotopes of Mg and Al) are potential participants in hydrostatic hydrogen burning. During the transformation of H into He in the CNO cycle, the relative abundances of the CNO nuclides vary, but their sum remains constant; i.e. the CNO nuclides play the role of catalysts. The same is true of the role of NeNa and MgAl nuclides in the NeNa and MgAl cycles. Detailed reaction schemes and the newest data on the reaction rates for all three cycles[6] can be found, for example, in the review by Arnould et al. [5]. Such approximate constancy of the C+N+O sum in spite of large differences in the individual abundances of C, N and O has, indeed, been observed in M13 and M3 by Smith et al. [178], in NGC 362 and NGC 288 by Dickenson et al. [70] and in ω Cen by Norris and Da Costa (NDC95). There is also some evidence for the constancy of the Mg+Al sum in M13 [107].

In contrast, the abundances of heavier α-process elements such as Si, Ca and Ti, which are thought to be formed in successive α-particle captures in massive stars and do not participate in hydrostatic hydrogen burning, do not display large scatters in globular-cluster red giants. On the contrary, their mean abundances are fairly consistent with the abundances of α elements in Population II field dwarfs, $\langle[\alpha/\text{Fe}]\rangle \approx +0.4$. The iron-peak elements Cr and Ni synthesised in supernova explosions likewise do not show abundance anomalies. In most globular clusters, the abundance of Fe itself remains surprisingly constant within a single cluster. An exception to this rule is ω Cen (and possibly M22), in which [Fe/H] varies from -1.8 to -0.8. Giants in ω Cen also show a growth in the abundances of s-process elements with [Fe/H], which undoubtedly indicates more complex chemical evolution (see NDC95).

Thus, the observations support the idea that the star-to-star variations in the surface elemental abundances (or at least an appreciable fraction of them) in globular clusters are most likely to have developed during hydrogen burning. The fact that there are fairly well-defined correlations between the overabundances of N, Na and Al on the one hand, and the underabundances of C, O and Mg on the other, clearly

---

[6] At the beginning of 1997.

DEEP MIXING IN GLOBULAR-CLUSTER RED GIANTS 99

testifies to a simultaneous origin for them. The next question we must answer is where these variations developed. In this section, we will attempt to provide an answer to this question. We will make use of the spectral analyses of the chemical compositions of red giants performed by Kraft et al. [107], Shetrone [170] and NDC95 for the globular clusters M13 and ω Cen. M13 is a very representative "normal" cluster (i.e. showing no peculiarities in its abundances of Fe or s-process elements), and makes a defining contribution to the so-called "global anticorrelation" of [O/Fe] with [Na/Fe] (in Kraft's terminology [103]). On the other hand, ω Cen provides evidence that the anticorrelation of O with Na and Al is not present only in "normal" clusters. Moreover, ω Cen is the only globular cluster that may not follow the usual global anticorrelation, since it has some giants with extremely high [Na/Fe] values (to 1!), much higher than the mean maximum ($\approx +0.5$) for "normal" clusters. Below, we will show how this anomaly can be explained theoretically.

Note that standard stellar-evolution theory is also not able to explain the low values of the ratio ^{12}C/^{13}C measured for Population I field giants with masses $M < 2\,M_\odot$ [84, 42]. Thus, in this sense, globular-cluster red giants are not unique. We emphasise, however, that the number of nuclides showing star-to-star abundance variations is greater in the latter sources than in Population I giants.

3.1.2 Computer Code Used

We used a modified version of the code described by Raffelt and Weiss [151] to calculate the evolution of a star with $M = 0.8\,M_\odot$ from the zero-age main sequence to the onset of the core helium flash. This modified version made use of new tables of opacity coefficients from [157, 95] and new rates of energy loss to neutrino radiation from [88]. The relative mass abundance of heavy elements was taken to be $Z = 5 \cdot 10^{-4}$ ($\lg Z/Z_\odot = -1.58$), fairly close to the metallicities of M13 ([Fe/H] $= -1.49$) and ω Cen ($-1.8 <$ [Fe/H] < -0.8). The initial helium abundance was taken to be $Y = 0.24$, in accordance with the chemical composition of matter formed during the Big Bang [208]. Our adopted value $\alpha = 1.48$ for the mixing length in the convective envelope (in units of the pressure scale height) was determined via calibration of our model for the Sun.

A small number of basis red giant models were selected from an evolutionary sequence. These were used to interpolate the temperature and density distributions in accordance with the procedure described in detail in [64] and in Section 2.1. Extra-mixing in the radiative zone between the hydrogen burning shell and the base of the convective envelope was modelled by diffusion. The results of these calculations of nucleosynthesis with extra-mixing depend on two parameters: the mixing depth δM_{mix}, defined as the relative mass co-ordinate of the deepest layer in the radiative zone to which the mixing extends, and the mixing rate, defined as the diffusion coefficient D_{mix} (δM is measured from the base of the hydrogen burning shell – more precisely, from the layer where the hydrogen abundance is $X = 10^{-4}$ – and is measured in fractions of the total mass of the radiative zone). The mixing rate cannot be arbitrarily high, since there exist "reasonable" estimates for its upper limit, namely $D_{\mathrm{mix}} \sim 10^7 - 10^9$ cm$^2 \cdot$s^{-1}(see [64]). The system of nuclear kinetics equations used in our mixing calculations includes all nuclides participating in reactions of the CNO, NeNa and MgAl cycles that are of interest for the problem at hand, as well as a small number of nuclides from pp-chain reactions (required to trace variations in the ^{3}He abundance). In total, the system takes into account 26 particles and 30 reactions.

We used the algorithm described in [68], which is similar to the scheme of Renzini and Voli [154], for the nucleosynthesis calculations for stars on the asymptotic giant branch. This algorithm takes into account nuclear reprocessing of stellar material in the hydrogen burning shell (in the AGB star!), at the base of the convective envelope between flashes of the helium burning shell ("hot-bottom burning"), and also in the convective layer formed during a helium flash. The stellar structure was described using a simple parametrization based, however, on data for precise models of AGB stars; the nucleosynthesis calculations were exact. Unfortunately, some of the model parameters remain poorly known. For example, the temperature at the base of the convective envelope between flashes and the fraction of reprocessed material in the convective helium shell that is carried to the surface by convection immediately after a flash cannot yet be determined with certainty [121]. For this reason, we concentrate in Section 3.1.4 only on our nucleosynthesis calculations for

DEEP MIXING IN GLOBULAR-CLUSTER RED GIANTS 101

intermediate-mass AGB stars, which depend only weakly on our choice of such parameters. As appropriate, we point out parameter values that lead to theoretical results in agreement with the observations. Unfortunately, our approach appears to be the only one possible at present, due primarily to uncertainties in our current understanding of convective overshoot in stars [78].

The system of nuclear kinetics equations we used in the AGB-star nucleosynthesis calculations take into account the same nuclides and reactions as in our calculations of diffusion mixing in red giants, as well as isotopes of Si; the 3α reaction; (α, γ) reactions with the participation of ^{12}C, ^{14}C, ^{14}N, ^{15}N, ^{16}O, ^{17}O, ^{18}O, ^{20}Ne, ^{21}Ne, ^{22}Ne, ^{24}Mg, ^{25}Mg and ^{26}Mg; (α,n) reactions with the participation of ^{13}C, ^{17}O, ^{18}O, ^{21}Ne, ^{22}Ne, ^{25}Mg and ^{26}Mg; neutron capture by the nuclides ^{12}C, ^{13}C, ^{16}O, ^{19}F, ^{20}Ne, ^{21}Ne, ^{22}Ne, ^{23}Na, ^{24}Mg, ^{25}Mg, ^{26}Mg, ^{27}Al, ^{28}Si, ^{29}Si, ^{30}Si and the mean heavy "nucleus" ^{31}X$_{14}$, which, to first approximation, replaces all nuclei heavier than ^{30}Si; and the reactions ^{14}C$(p,\gamma)^{15}$N, ^{14}N$(n,p)^{14}$C, ^{19}F$(\alpha,p)^{22}$Ne, ^{26}Al$^{\rm g}$$(n,p)^{26}$Mg, ^{21}Ne$(n,\alpha)^{18}$O and ^{26}Al$^{\rm g}$$(n,\alpha)^{23}$Na. The total numbers of nuclides and reactions included in this system of equations are 26 and 69.

To elucidate the degree to which s-process nucleosynthesis influences the abundances of heavy elements supplied to the interstellar medium of the globular cluster by its intermediate-mass (and low-metallicity) AGB stars, we prepared a special programme taking into account all the reactions listed above together with slow neutron capture by numerous heavy nuclides; the resulting system of nuclear kinetics equations was then solved for 409 particles coupled by 1273 reactions. We assumed that the neutrons have equilibrium local concentrations in each of the mass zones into which the helium burning shell is divided.

Naturally, we used the same reaction rate data for all the nucleosynthesis calculations. For reactions involving charged particles, we usually adopted the corresponding data from the tables of Caughlan and Fowler [36]. Exceptions are the reactions ^{17}O$(p,\alpha)^{14}$N and ^{17}O$(p,\gamma)^{18}$F, for which we used the rates presented by Landré et al. [112] with the uncertainty factors $f_1 = 0.2$ and $f_2 = 0.1$ recommended in [16]. In accordance with the suggestion of Woosley and Weaver [217], we increased the rate of the reaction ^{12}C$(\alpha,\gamma)^{16}$O from [36] by a factor of 1.7. We also adopted the reaction rates for the NeNa cycle presented by El Eid and Champagne [72] as a test, but did not include these in our main calculations (for reasons described in Section 3.1.3). Below, we specifically indicate places where the NeNa-cycle reaction rates of [72] were used.

The cross-sections for neutron capture were taken from the reviews [76, 93, 218, 6, 152, 8, 50, 164]. The rates of β decays were interpolated in temperature and density using the tables of [193].

Finally, we used a modified version of the code of Paczynski [141], described in [57], to calculate the structure of metal-poor, massive stars on the zero-age main sequence (Section 3.1.4).

3.1.3 Deep-Mixing Scenarios

By the end of the 1970s, the existence of star-to-star abundance variations in globular clusters was well established for C and, to a lesser extent, N [9, 69, 138, 53], while no quantitative information about such variations was available about O. In 1979, Sweigart and Mengel [192] found that in their models for metal-poor red giants, the layer in which C begins to be transformed into N (the "C layer") and, at very low metallicities, even the layer in which O is burned to form N (the "O layer") are fairly far from the bulk of the hydrogen burning shell where H is transformed into He. This meant that meridional circulation flows induced by the star's rotation could come rather close to the hydrogen burning shell, and begin to carry outward material depleted in C and O and enriched in N. Usually, a molecular weight gradient forms a barrier that cannot be overcome by meridional circulation, however this occurs only where there are appreciable changes in the H abundance. The further development of Sweigart and Mengel's idea became known as the "deep(extra)-mixing scenario" or simply the "evolutionary scenario". Often, the nature of this extra-mixing was not discussed. Exceptions were the pioneering work of Sweigart and Mengel [192] and the related work of Smith and Tout [177], in which it was shown that meridional circulation can provide the required mixing rate, even in a classical description, and the work of Charbonnel [43], who used simple estimates for the mixing rate based on the more complex model of Zahn [219], taking into account the interaction between meridional circulation and turbulent diffusion. Like this section, other studies concentrated primarily on nucleosynthesis calculations, addressing the question of whether or not extra-mixing (whatever its origin) is capable of explaining the entire spectrum of abundance anomalies observed in globular-cluster stars (and correlations between them).

In 1980, following the first reports of Cottrell and Da Costa [49] and Norris et al. [137] that overabundances of N were accompanied by enhanced abundances of Na and Al in NGC 6752, data began to be accumulated testifying that this was a general property of many globular clusters. Moreover, in 1989, Paltoglou and Norris [142] obtained data hinting at the existence of an anticorrelation between the abundances of Na and O in ω Cen, heralding the later discovery of the global anticorrelation between [O/Fe] and [Na/Fe] by Kraft et al. [106]. Extra-mixing scenarios began to lose support, since it was not at all understood at that time how nuclei like Na and Al – with rather high charges and therefore high Coulomb barriers to nuclear reactions with charged particles – could be synthesised during hydrogen burning in low-mass stars.

Extra-mixing was "rehabilitated" by evidence for the existence of low-energy resonances in the reactions ^{22}Ne(p,γ)^{23}Na, ^{25}Mg(p,γ)^{26}Al and ^{26}Mg(p,γ)^{27}Al. First, Denissenkov and Denissenkova [61] demonstrated that the temperature near the O layer was sufficiently high for the synthesis of sodium to occur via the reaction ^{22}Ne(p,γ)^{23}Na (thanks to resonance!), even more rapidly than the reaction ^{16}O(p,γ)^{17}F, which is responsible for O burning. Further, Langer et al. [115] found that ^{27}Al could also be synthesised (primarily from ^{25}Mg) slightly deeper than the O layer in metal-poor red giants. These results are illustrated in Fig. 3.1, which depicts abundance profiles for several nuclides participating in the CNO, NeNa and MgAl cycles in the part of the radiative zone adjacent to the hydrogen burning shell in a stellar model with $M = 0.8\,M_\odot$, luminosity $\lg(L/L_\odot) = 3.0$ and $Z = 5\cdot 10^{-4}$. The vertical line segments along the horizontal axis correspond to the locations of layers in which the hydrogen abundances are lower (as a consequence of nuclear reactions) by 1, 5 and 10% compared to the atmospheric abundance (from right to left). When modelling extra-mixing, we did not assume that it penetrated very deeply into the hydrogen burning shell. This was motivated by the following circumstances: *(i)* it follows from theoretical considerations that the fairly large molecular-weight

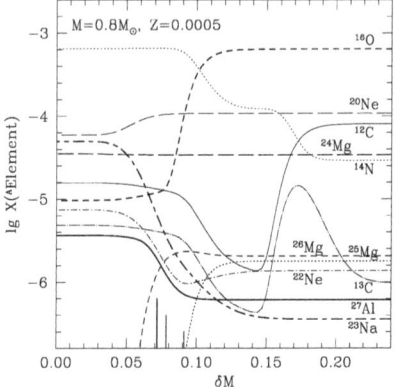

Figure 3.1. Abundance profiles for some nuclides participating in the CNO, NeNa and MgAl cycles in the radiative zone, adjacent to the hydrogen burning shell, in a red giant model with mass $M = 0.8\,M_\odot$, luminosity $\lg(L/L_\odot) = 3.0$ and heavy element content $Z = 5 \cdot 10^{-4}$ (a fairly close approximation to the metallicities of the globular clusters M13 and ω Cen). The relative mass co-ordinate δM is measured from the lower boundary of the hydrogen burning shell (where $X = 10^{-4}$) and is given in fractions of the mass contained between the hydrogen burning shell and the base of the convective envelope. The vertical line segments along the horizontal axis show the locations of layers in which the hydrogen abundance has decreased from the atmospheric value by 1, 5 and 10% (from right to left).

gradient should hinder the action of both meridional circulation and turbulent diffusion (see, for example, [101, 195]); *(ii)* the observed Na overabundances also place constraints on the depth (and rate) of the extra-mixing (see below); *(iii)* our calculation algorithm from [64] did not allow extra-mixing to carry too much He to the surface, since otherwise it would be necessary to take into account the feedback effect of the mixing on the internal structure and evolution of the red giants.

Nucleosynthesis with extra-mixing in globular-cluster red giants has also been discussed by Cavallo *et al.* [38] and Langer *et al.* [116]. As in our analysis, the papers of Cavallo *et al.* use exact stellar models to determine the physical conditions under which nuclear reactions proceed, while this is not true of the calculations of Langer *et al.* The results of these studies supplement our own investigations, since they consider the influence of metallicity on the efficiency of deep mixing.

Initial chemical composition In the calculations of the abundance profiles presented in Fig. 3.1, the initial chemical composition was determined as follows. First, solar abundances (from [2]) for all nuclides relevant to the problem in hand were multiplied by $Z/Z_\odot < 1$. Further, the resulting abundances of α elements (^{16}O, ^{20}Ne, ^{24}Mg and ^{28}Si) were increased by a factor of 2.5, so that their mean $\langle[\alpha/\text{Fe}]\rangle \approx +0.4$ corresponded to the observed abundances for Population II field dwarfs (see Wheeler *et al.* [215]). Next, the abundances of Na and Al were reduced to $[\text{Na/Fe}] = [\text{Al/Fe}] = -0.4$ to take into account the known effect that the abundances of nuclides with odd mass numbers are usually lower than those of nuclides with even mass numbers [215]. Finally, to correctly reproduce the global anticorrelation of [O/Fe] with [Na/Fe], the initial abundance of ^{22}Ne was reduced to $[^{22}\text{Ne/Na}] = 0$.

We did not introduce any corrections to take into account variations in the surface chemical composition during the first dredge-up, since these are negligible compared to the variations produced by extra-mixing. In concluding this subsection, we point out that, in accordance with the above algorithm, the initial isotopic composition of magnesium in our calculations was $^{24}\text{Mg}/^{25}\text{Mg}/^{26}\text{Mg} = 90.5/4.5/5.0$, and not 79/10/11, as for the Sun.

Abundances of CNO elements and Na The following three features in Fig. 3.1 merit special attention. *(i)* Closer to the hydrogen burning shell, the Al abundance increases much more than the Na abundance. There are two steps in the Na-abundance profile, the first due to burning of ^{22}Ne and the second due to partial burning of the much more abundant ^{20}Ne. *(ii)* The abundance of ^{24}Mg does not change at all. We should emphasise that, even if we take a model for a star immediately before the core helium flash, the ^{24}Mg abundance is lowered only very slightly, and then only within the hydrogen burning shell. *(iii)* The abundance of ^{12}C first decreases with depth, then increases after the CN cycle has reached an equilibrium regime, since the abundance of ^{12}C (and ^{13}C) then varies in proportion to the

^{14}N abundance, which increases with depth. We therefore expect that when the depth (and rate) of mixing are varied, the surface abundances of ^{12}C and ^{13}C will behave qualitatively differently from, say, the O abundance. For example, the surface abundance of C may be *higher* for a star with deep mixing ($\delta M_{\mathrm{mix}} \approx 0.06$) than for another with less deep mixing ($\delta M_{\mathrm{mix}} \approx 0.13$).

We found that extra-mixing with depth and rate $\delta M_{\mathrm{mix}} = 0.05$ and $D_{\mathrm{mix}} = 5 \cdot 10^8$ cm$^2 \cdot$s^{-1} is sufficient to reproduce well all known observational correlations between surface abundances except for the relationship between [O/Fe] and [Al/Fe] in red giants of ω Cen (solid curves in Fig. 3.2). The observational data for Fig. 3.2 were taken from NDC95. The N abundances were refined by adding corrections with a mean value of $+0.5$ to all [N/Fe] values, in order to achieve consistency with the data of Brown and Wallerstein [22] (Norris and Da Costa, private communication, 1997).

Of course, we cannot assume that all low-mass stars in ω Cen (or in any other cluster) had identical chemical compositions (for example, close to that described in the preceding section) before they became red giants; for example, the surfaces of stars in close binary systems could be "contaminated" by products of nucleosynthesis produced in more massive, and therefore more evolved, companions. Detailed differences in [C/Fe] are indeed observed among giants in ω Cen (see the group of stars with strong molecular CO bands in Fig. 3.2). Nevertheless, our calculations assumed that the variations in the initial abundances for most of the studied stars were small.

To explain the observed dependence of [O/Fe] on [Na/Fe] for giants in M13, which forms the "skeleton" of the global anticorrelation between [O/Fe] and [Na/Fe] for "normal" globular clusters, we had to consider somewhat less deep and simultaneously more rapid mixing than in the case of ω Cen: $\delta M_{\mathrm{mix}} = 0.06$, $D_{\mathrm{mix}} = 2.5 \cdot 10^9$ cm$^2 \cdot$s^{-1}. The true nature of this difference in the diffusion coefficients will not be understood until the intrinsic physical mechanism for the extra-mixing is established. The results of our calculations for M13 are shown by the dashed curves in Fig. 3.3, for which the observed Na and Al abundances were taken from [107] and corrected by $+0.05$ and -0.25 respectively to compensate for the differences in oscillator strengths used in [107] and NDC95. Note that the agreement between the observed and theoretical dependences of [O/Fe] on [Al/Fe]

DEEP MIXING IN GLOBULAR-CLUSTER RED GIANTS 107

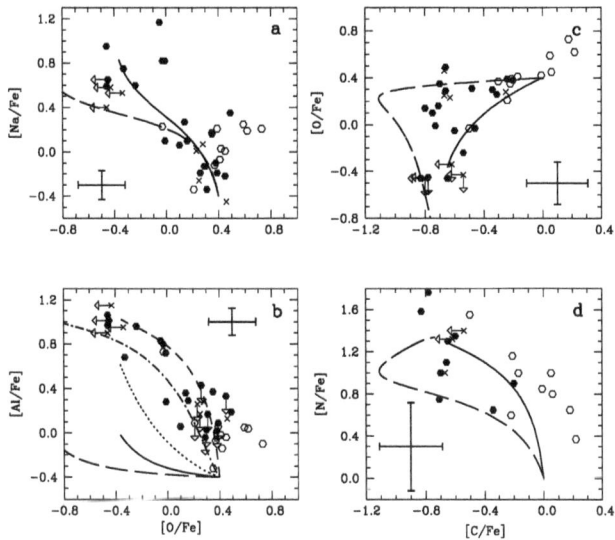

Figure 3.2. Correlations between various elemental abundances observed for red giants of ω Cen (the symbols show data from [139]), together with the results of our calculations of nucleosynthesis with deep extra-mixing for two sets of depths and rates (diffusion coefficients) for the mixing $(\delta M_{\rm mix}; D_{\rm mix}, {\rm cm}^2 \cdot {\rm s}^{-1})$: $(0.05; 5 \cdot 10^8)$ – solid, dotted and short-dashed curves, $(0.06; 2.5 \cdot 10^9)$ – long-dashed and dot–dashed curves. The first set provides a closer agreement with the observations for ω Cen, while the second better reproduces the anticorrelation between [O/Fe] and [Na/Fe] in M13 (Fig. 3.3). The dotted curve in panel **b** was calculated with initial abundance $[^{25}{\rm Mg/Fe}] = 1.2$, while the short-dashed and dot–dashed curves were obtained assuming $[^{25}{\rm Mg/Fe}] = 1.1$ and a rate for the reaction $^{26}{\rm Al}^{\rm g}({\rm p},\gamma)^{27}{\rm Si}$ a factor of 10^3 higher than the corresponding values from [36] (see Section 3.1.3 for further discussion). The white and black symbols denote stars with strong and weak molecular CO bands, and the crosses represent stars with unknown CO-band intensities, according to the data of [139]. In panel **d**, all the N abundances from [139] were shifted by on average $+0.5$ dex in order to achieve consistency with the data of Brown and Wallerstein [22]. The mean observational errors are indicated (large crosses).

is poor. Note also that even for mixing depths $\delta M_{\mathrm{mix}} \approx 0.05 - 0.06$ (see Fig. 3.1), the chosen diffusion rate is such that the additional amount of hydrogen reprocessed by the hydrogen burning shell (compared to the standard model) is modest. In the two cases considered above (ω Cen and M13), the final surface abundances of H decrease by only 9.9 and 7.5% respectively, and, as shown by Sweigart [191], the course of the star's evolution on the red giant branch is virtually unaffected.

By way of comparison, the solution providing the best agreement with the global anticorrelation (for giants in M13) is also presented in Fig. 3.2 (long-dashed curve). We can conclude from Fig. 3.2a that the cluster ω Cen may contain red giants in whose atmospheres we see Na produced not only from ^{22}Ne, but also from ^{20}Ne. In other words, the extra-mixing in some giants in ω Cen may reach even the second (deeper) step in the Na-abundance profile (see Fig. 3.1). To our knowledge, ω Cen is the only globular cluster studied to date that displays this property.

A more careful analysis of the differences between ω Cen and M13 shows that they are associated with metal-rich stars in ω Cen. We will discuss this in more detail in Section 3.1.3 below. Here, it suffices to note that our conclusion that the mixing is deeper in ω Cen giants with unusually high Na abundances (these same stars also have higher values of [Fe/H]) remains valid, since the regions in which Na is synthesised are closer to the hydrogen burning shell in giants with higher metallicity so that deeper (and more rapid) mixing is required to reach these layers (as confirmed by the calculations of Cavallo et al. [38]).

As already noted above, a detailed analysis of structural and evolutionary differences in parameters important for deep-mixing scenarios for models of red giants with various metallicities has been carried out by Cavallo et al. [38]. Their main conclusions with regard to the effects of metallicity can be briefly summarised as follows. *(i)* The hydrogen burning shell moves outward more rapidly as metallicity grows, giving less time for nuclear reprocessing of the material directly beneath the shell. *(ii)* In more metal-rich models, layers with temperatures sufficiently high to affect the elemental abundances in the envelope lie closer to the hydrogen burning shell (a similar conclusion was drawn in [192]), while the temperature of these layers is

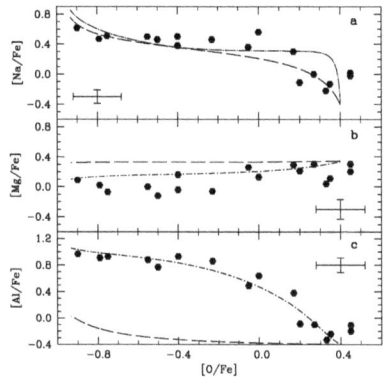

Figure 3.3. Anticorrelation of [O/Fe] with [Na/Fe] and [Al/Fe] and correlation of [O/Fe] with [Mg/Fe] observed for giants in M13 (symbols) compared with the results of our calculations for $\delta M_{\mathrm{mix}} = 0.06$ and $D_{\mathrm{mix}} = 2.5 \cdot 10^9$ cm$^2 \cdot$s^{-1}. The observational data are taken from Kraft et al. [107], with corrections of +0.05 and −0.25 added to the [Na/Fe] and [Al/Fe] values to take into account the differences in the oscillator strengths used by Kraft et al. [107] and by Norris and Da Costa [139]. The dashed curves were calculated using standard input physics (see text). The dot–long-dashed curve in panel **a** was obtained using new data on the rates of NeNa-cycle reactions from [72], while the dot–short-dashed curve (panels **b** and **c**) was obtained using the initial values [^{24}Mg/Fe] = 0 (the adopted standard value was +0.4) and [^{25}Mg/Fe] = 1.1 and a rate for the reaction ^{26}Al^{g}(p,γ)^{27}Si a factor of 10^3 higher than the rates given in [36]. The observational errors are indicated (large crosses).

lower than in metal-poor models. These facts led Cavallo et al. [38] to conclude that variations in surface abundances should be greater and occur for a larger number of elements in red giants with low metallicities than in those with higher metallicities (of course, in the presence of extra-mixing). In contradiction with these theoretical predictions, metal-rich giants in ω Cen have higher Na excesses than their metal-poor neighbours in the cluster (see NDC95 and Section 3.1.3 below). However, it is too early to consider this a serious disagreement between the theory and observations, since selection effects play a major role in the study of metal-rich giants carried out in NDC95. In addition, as already noted, the massive cluster ω Cen may well have a unique chemical evolution history.

Further analyses (as free as possible of selection effects) of abundances in giants in ω Cen, whose metallicities vary over a rather wide interval, would be very useful in the search for observational evidence for metallicity dependences of various extra-mixing parameters. Unfortunately, comparing the abundances for red giants from different clusters cannot resolve this question, since it is likely that the efficiency of mixing varies from cluster to cluster.

It was suggested in NDC95 that there was a peculiar saturation in the C abundances of giants in ω Cen at [C/Fe] $\approx$ −0.8. The existence of this limiting value finds a natural explanation in our calculations: deep mixing (as in the case of ω Cen) reaching layers in which the C abundance begins to increase with depth (in Fig. 3.1, this occurs at $\delta M_{\mathrm{mix}} < 0.15$; see comment *(iii)* at the beginning of this section) cannot lead to a large underabundance of C at the stellar surface (see solid and dashed curves in Fig. 3.2c).

Calculations with deep extra-mixing for giants in ω Cen show that during the last 50% of the time from the onset of mixing to the core helium flash, the ratio ^{12}C/^{13}C at the stellar surface gradually decreases from 9 to 6, in close agreement with the values 4–6 measured by Brown and Wallerstein [21] for stars near the tip of the red giant branch.

The dot–long-dashed curve in Fig. 3.3a shows the results of our calculations with deep mixing using the new rates for NeNa-cycle reactions of El Eid and Champagne [72]. The shape of this dependence differs appreciably from that of the long-dashed curve, calculated using the reaction rates of [36], especially for $0 \leq$ [O/Fe] $\leq +0.4$. This difference is entirely due to the substantial growth in the rate of the reaction ^{22}Ne(p,γ)^{23}Na, which shifts the first step in the Na-abundance profile beyond the O layer (Fig. 3.4). As a result, the deep mixing first leads to a rapid growth in the surface abundance of Na, only after which does the O abundance begin to decrease. Unfortunately, errors in the spectral analysis of abundances prevent us from being able to make a definite choice between the old and new reaction rates based on Fig. 3.3a.

DEEP MIXING IN GLOBULAR-CLUSTER RED GIANTS 111

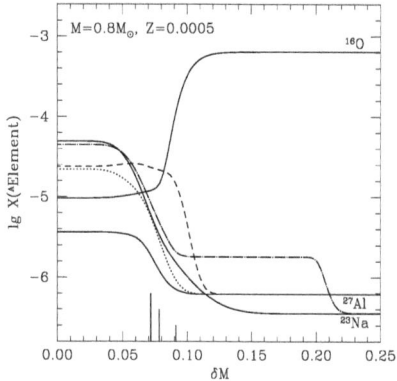

Figure 3.4. Distribution of O, Na and Al abundances near the hydrogen burning shell calculated under various assumptions: standard assumptions (see text) – solid curve; new NeNa-cycle reaction rates from [72] – dot–long-dashed curve; initial abundance [^{25}Mg/Fe] = 1.2 – dotted curve; [^{25}Mg/Fe] = 1.1 and rate of the reaction ^{26}Al^{g}(p,γ)^{27}Si increased by a factor of 10^3 over the value of [36] – short-dashed curve.

Thus, a deep-mixing scenario applied to the globular cluster ω Cen is able to reproduce the observed variations of the C, N, O and Na abundances (and the ratio ^{12}C/^{13}C). At the same time, it cannot explain the anticorrelation of [O/Fe] with [Al/Fe] (Fig. 3.2b, solid curve and Fig. 3.3c, dashed curve) or the correlation between [O/Fe] and [Mg/Fe] (Fig. 3.3b, dashed curve).

When does the deep mixing begin? In the cases considered above (M13 and ω Cen), the deep-mixing calculations began with a red giant model in which the hydrogen burning shell had just crossed the jump in the H abundance left by the base of the convective envelope at the end of the first dredge-up. In this and several neighbouring models, the luminosity is slightly diminished, due to the adjustment of the hydrogen burning shell to the growth in the amount of hydrogen fuel. As a result, the course of the star's evolution slows near this place, and the luminosity function of a globular cluster will display a local maximum at the corresponding magnitude. Sweigart and Mengel [192] were the first to propose that extra-mixing in red giants begins precisely at this point in the star's evolutionary sequence, since in all preceding models there is a molecular-weight gradient between the hydrogen burning shell and the jump in the H abundance, which arises already in the MS phase, preventing the action of mixing (meridional circulation). Observations of anomalous evolutionary variations in the ratio $^{12}C/^{13}C$ and the ^{3}He and ^{7}Li abundances in field giants confirm this conclusion (see [42, 43, 45]). However, this is in contradiction with other observational data showing, for example, that the C abundance in the atmospheres of stars in M92 (and a number of other clusters) begins to decrease much earlier than proposed by Sweigart and Mengel (see [117]). Moreover, the spectral analysis of M13 red giants carried out by Kraft *et al.* [107] revealed stars with overabundances of Na and Al and underabundances of O (and even Mg!) in the presence of fairly low luminosities. Thus, at least in the stars in M13, deep extra-mixing begins earlier than Sweigart and Mengel thought (assuming, of course, that it is in fact extra-mixing that is responsible for the observed abundance variations). Our opinion on this is as follows. Analysis of the chemical structure of our models shows that the molecular weight gradient that arises during the MS phase is actually appreciably lower than the gradient at the point reached by the extra-mixing in calculations that can reproduce the observed Na and Al overabundances in globular-cluster giants (using the reaction rates of [36]). We therefore conclude that in any globular cluster containing giants with excesses of Al (and Na) and underabundances of O, deep mixing (even if we don't know its precise nature) begins to act long before the time suggested by Sweigart and Mengel.

A potentially more serious problem for extra-mixing could be its suppression by the jump in the H abundance, near which in standard models the molecular weight experiences variations within a fairly narrow range of mass co-ordinates. Recall that this abundance jump is created by the base of the convective envelope when it achieves its greatest depth. Since manifestations of extra-mixing such as an evolutionary decrease in the C abundance are visible long before this jump is engulfed by the hydrogen burning shell (in M92, for example [117]), we are forced to suppose that extra-mixing, possibly acting

DEEP MIXING IN GLOBULAR-CLUSTER RED GIANTS   113

jointly with convective overshooting at the base of the envelope, prevents the formation of the H jump (or smooths it). On the other hand, since a local maximum in luminosity is observed for many globular clusters (see [1]), it is possible that the H jumps in the red giants of these clusters have not all been smoothed.

Mg and Al abundances   As we can see in Fig. 3.1, it is impossible to synthesise Al from ^{24}Mg in standard evolutionary calculations (see also comment *(ii)* in connection with Fig. 3.1 in Section 3.1.3). At the same time, Al can be produced from ^{25}Mg (Fig. 3.1), but in quantities which are insufficient to explain the observational data for ω Cen (see Fig. 3.2b, solid curve) and M13 (Fig. 3.3c, dashed curve). Some time ago, it was possible to speculate that the initial abundance of ^{25}Mg was anomalously high in some globular clusters compared to the value obtained using the algorithm described in Section 3.1.3 (see [114, 64]). In the solar chemical composition, ^{24}Mg is the most abundant isotope of magnesium: ^{24}Mg/^{25}Mg/^{26}Mg $= 79/10/11$ [2]. If we suppose that in globular-cluster red giants, initially $[^{25}\text{Mg/Fe}] > 1.0$, we can obtain the observed overabundances of Al (see below). In this case, however, due to the appreciably growing contribution of ^{25}Mg in the sum Mg $= {}^{24}$Mg$+{}^{25}$Mg$+{}^{26}$Mg and the evolutionary transformation of this ^{25}Mg into Al, we expect a decrease in the total magnesium content [Mg/Fe] with increasing [Al/Fe]. Such a dependence has, indeed, been found for giants in M13 (Fig. 3.3b, symbols) by Shetrone [169] and Kraft *et al.* [107]. On the other hand, an analysis of the isotopic composition of magnesium in a sample of six M13 giants [170, 191] unexpectedly showed that the stars with the highest values of [Al/Fe] had significant underabundances of ^{24}Mg, and not of ^{25}Mg (in the five stars with the highest [Al/Fe] values, on average, $\langle[^{24}\text{Mg/Fe}]\rangle = -0.33$). This same analysis revealed (for the first time in the entire history of studies of stellar chemical compositions) anomalous magnesium isotopic ratios, with an increased contribution of the sum ^{25}Mg$+^{26}$Mg to the total Mg content (the $[(^{25}\text{Mg}+^{26}\text{Mg})/\text{Fe}]$ values reach $+0.21$, and the mean isotopic ratios $\langle^{24}\text{Mg}\rangle/\langle^{25}\text{Mg}\rangle/\langle^{26}\text{Mg}\rangle = 56/22/22$). Unfortunately, Shetrone [170] could not separate the isotopes ^{25}Mg and ^{26}Mg, and could obtain only their combined content, so that he simply assumed that ^{25}Mg and ^{26}Mg had the same abundances in his Mg isotope ratio estimates.

The simplest solution to the problem of the origin of the anticorrelation of [Mg/Fe] with [Al/Fe] in giants in M13 would be the discovery of a strong, but as yet unknown, low-energy resonance in the reaction ^{24}Mg(p,γ)^{25}Al; this would simultaneously provide an explanation for the anticorrelations between O and Al in M13 and ω Cen and take into account the analysis of the isotopic abundance of Mg by Shetrone [170]. The new rate for this reaction should be comparable to that of the reaction ^{25}Mg(p,γ)^{26}Al if the extra-mixing is to reach depths at which the abundance of ^{24}Mg is reduced (see the ^{25}Mg profile in Fig. 3.1). In this case, Al would be produced in the reaction chain ^{24}Mg(p,γ)^{25}Al($\beta^+\nu$)^{25}Mg(p,γ)^{26}Al^g(p,γ)^{27}Si($\beta^+\nu$)^{27}Al, while the β decay ^{26}Al^g($\beta^+\nu$)^{26}Mg and the channel ^{25}Mg(p,γ)^{26}Al^m($\beta^+\nu$)^{26}Mg would provide a sufficiently large final total ^{25}Mg+^{26}Mg, where, in this case, the main contribution is made by the isotope ^{25}Mg. "Unfortunately", nuclear physicists have no doubt about the accuracy of existing data on the rate of the reaction ^{24}Mg(p,γ)^{25}Al (see [5, 222]).

Note that uncertainty in the rate of NeNa-cycle reactions also influences the resulting magnesium isotopic abundances, since the NeNa cycle produces ^{24}Mg, with the reaction rates of El Eid and Champagne [72] making this production more powerful. However, it becomes even more difficult to reduce the ^{24}Mg abundance near the hydrogen burning shell when using the reaction rates of El Eid and Champagne [72], which, in turn, makes it more problematic to explain the anticorrelation between [Mg/Fe] and [Al/Fe] in the framework of extra-mixing scenarios.

A possible alternative is that in globular-cluster red giants with excesses of Na and Al and deficits of O and Mg, we actually observe the products of hydrogen burning at higher temperatures (say, $T_6 \equiv T/10^6$ K ≈ 70) than the temperature that is reached in the hydrogen burning shell ($T_6 \leq 55$). We will discuss this possibility in the next section.

If the two suggestions expressed above are not realised, a third and final possibility is that unusually high overabundances of Al in globular-cluster red giants are a consequence of an excess initial abundance of ^{25}Mg. However, if we now wish to avoid a conflict with the Mg isotopic abundance analysis of Shetrone [170], we must also explain how a deficit of ^{24}Mg develops in stars with especially high initial abundances of ^{25}Mg. In Section 3.1.4, we consider primordial scenarios in globular clusters that are potentially able to create the required isotopic mixture of Mg, and until then, we will simply assume that some low-mass stars in M13 and ω Cen had initial overabundances of ^{25}Mg.

The dotted curve in Fig. 3.2b (see also Fig. 3.4) shows the variations of [Al/Fe] with [O/Fe] at the surface of a red giant with initial composition $[^{25}\text{Mg/Fe}] = 1.2$. Here, we used the same values of δM_{mix} and D_{mix} as earlier, since these precise values fit well three other correlations in ω Cen. Unfortunately, as the figure shows, the results of these new calculations poorly reproduce the anticorrelation between [O/Fe] and [Al/Fe] in ω Cen, with the disagreement being both quantitative (not enough Al is produced) and qualitative (the dotted curve has a different shape than the observations). The quantitative difference is due to the fact that for the reaction rates of [36], the channel $^{25}\text{Mg}(p,\gamma)^{26}\text{Al}(\beta^+\nu)^{26}\text{Mg}$ dominates over the channel $^{25}\text{Mg}(p,\gamma)^{26}\text{Al}^{\text{g}}(p,\gamma)^{27}\text{Si}(\beta^+\nu)^{27}\text{Al}$ that produces Al, so that a large fraction of the ^{25}Mg is spent on ^{26}Mg instead of ^{27}Al. The wrong shape of the theoretical dependence of [Al/Fe] on [O/Fe] comes from the need to wait until enough $^{26}\text{Al}^{\text{g}}$ has accumulated for the reaction chain $^{26}\text{Al}^{\text{g}}(p,\gamma)^{27}\text{Si}(\beta^+\nu)^{27}\text{Al}$ to begin to have a rate competitive with the β decay $^{26}\text{Al}^{\text{g}}(\beta^+\nu)^{26}\text{Mg}$. The undesirable action of these effects can be reduced by increasing the rate of the reaction $^{26}\text{Al}^{\text{g}}(p,\gamma)^{27}\text{Si}$. In principle, this may be possible, since, according to Arnould et al. [5], the rate of the reaction $^{26}\text{Al}^{\text{g}}(p,\gamma)^{27}\text{Si}$ in [36] may be underestimated by a factor of up to 10^3 in the temperature range corresponding to hydrogen shell burning ($T_6 \approx 40-55$). If we use this uncertainty factor, the channel for Al synthesis becomes so broad that we must even adopt a slightly lower value for the initial abundance of ^{25}Mg. The short-dashed curve in Fig. 3.2b (see also Fig. 3.4) was calculated with $[^{25}\text{Mg/Fe}] = 1.1$ and a rate for the reaction $^{26}\text{Al}^{\text{g}}(p,\gamma)^{27}\text{Si}$ a factor of 10^3 higher than in [36]. The dot–short-dashed curve in Figs. 3.2b and 3.3c were obtained for the same conditions but with a depth and rate of extra-mixing corresponding to the anticorrelation between the O and Na abundances in M13. We can see that the theory and observations are now in agreement! The initial abundance of ^{24}Mg does not influence the results of these calculations at all. The dot–long-dashed curve in Fig. 3.3b was obtained assuming an initial abundance $[^{24}\text{Mg/Fe}] = 0$ (instead of +0.4), which, as expected, leads to an appreciable drop in the total magnesium abundance [Mg/Fe] with growth in [Al/Fe]. However, it appears that the observations require an even greater deficit of the initial ^{24}Mg abundance.

If we adopt [^{24}Mg/Fe] = 0, [^{25}Mg/Fe] = 1.1 and [^{26}Mg/Fe] = 0, this will be equivalent to the values [(^{25}Mg+^{26}Mg)/Fe] = 0.81, [Mg/Fe] = 0.33 and ^{24}Mg/^{25}Mg/^{26}Mg = 37/58/5, i.e. a mixture of Mg isotopes in which ^{25}Mg dominates. Further, for extra-mixing parameters corresponding to M13 giants (δM_{mix} = 0.06, D_{mix} = $2.5 \cdot 10^9$ cm$^2 \cdot$s^{-1}), we find immediately preceding the core helium flash [^{24}Mg/Fe] = 0.004, [^{25}Mg/Fe] = −0.86, [^{26}Mg/Fe] = 0.61, and consequently [(^{25}Mg+^{26}Mg)/Fe] = 0.35, [Mg/Fe] = 0.10 and ^{24}Mg/^{25}Mg/^{26}Mg = 64/1/35. However, since the observations are not able to separate ^{25}Mg from ^{26}Mg, this final mixture of magnesium isotopes does not differ from ^{24}Mg/^{25}Mg/^{26}Mg = 64/18/18, which is close to the mean values observed by Shetrone [170].

In his analysis of the Mg isotopic mixture, Shetrone considered the abundances of ^{25}Mg and ^{26}Mg to remain constant in the course of extra-mixing in red giants. However, as we have seen, there are at least two alternative interpretations.

(i) The isotopic mixture for Mg was initially composed almost entirely of ^{24}Mg (as, for example, in the giant L 598 from Shetrone's sample, which has ^{24}Mg/^{25}Mg/^{26}Mg = 94/3/3 and for which there are no signs of extra-mixing). If there is also a low-energy resonance in the reaction ^{24}Mg(p,γ)^{25}Al, mixing will decrease the ^{24}Mg abundance, while the ^{25}Mg+^{26}Mg and ^{27}Al abundances will increase.

(ii) The initial abundance of ^{24}Mg was low, while that of ^{25}Mg, on the contrary, was high (as a result of primordial nucleosynthesis in the cluster). During deep mixing, the ^{24}Mg abundance remains constant, while the amount of ^{27}Al grows due to burning of ^{25}Mg. The isotope ^{26}Mg, which is also synthesised from ^{25}Mg, begins to dominate in the sum ^{25}Mg+^{26}Mg.

Both cases will lead to approximately the results observed by Shetrone.

Observational properties in ω Cen as functions of metallicity In Section 3.1.3 we recalled that it is the more metal-rich stars in ω Cen that are primarily responsible for the observational differences in the surface abundances of elements in red giants in M13 and ω Cen. In Fig. 3.5, we have separated the sample of stars from ω Cen studied in NDC95 into two groups: to the left are objects with [Fe/H] < −1.3 and in the centre are those with [Fe/H] > −1.3. For comparison, Shetrone's data [169] for the more metal-rich clusters M5 ([Fe/H] = −1.2) and M71 ([Fe/H] = −0.8) are shown to the right. Figure 3.5 also shows the results of our calculations that best reproduce the anticorrelation of [O/Fe] with [Na/Fe] in M13 (dashed curve, taken from Fig. 3.3a).

Analysis of Fig. 3.5 reveals several interesting features. First, we can see that the anticorrelation between [O/Fe] and [Na/Fe] for less metal-rich stars in ω Cen is similar to that observed in M13. On the contrary, metal-rich stars in ω Cen have values of [Na/Fe] that exceed those for stars in M13 by about 0.4. Such large overabundances of Na are absent from the samples for M5 and M71, although these latter clusters have similar metallicities.

Second, the same anticorrelation of [O/Fe] with [Mg/Fe] established by Shetrone [169] and Kraft et al. [107] for M13 is visible in the group of metal-poor stars in ω Cen. The scatter in [Mg/Fe] ($\approx 0.5 - 0.6$) is larger than for M13 (≈ 0.3, see Fig. 3.3), possibly reflecting the presence of stronger mixing in ω Cen. However, this is very different from the behaviour of the dependence of [Mg/Fe] on [O/Fe] for metal-rich stars in ω Cen, which show very little scatter in [Mg/Fe].

Finally, both groups of stars in ω Cen apparently follow the same dependence of [Al/Fe] on [O/Fe]. An important circumstance accompanying this fact is the almost total absence of any dependence of [Mg/Fe] on [Al/Fe] in the metal-rich ω Cen giants, in strong contrast to the observations for metal-poor giants in ω Cen and M13!

In summary, compared to their metal-poor neighbours, metal-rich giants in ω Cen have similar Al overabundances, higher Na overabundances and normal Mg abundances (without deficits). Thus, these metal-rich stars in ω Cen provide important information for future modelling of the processes leading to peculiarities in their chemical compositions.

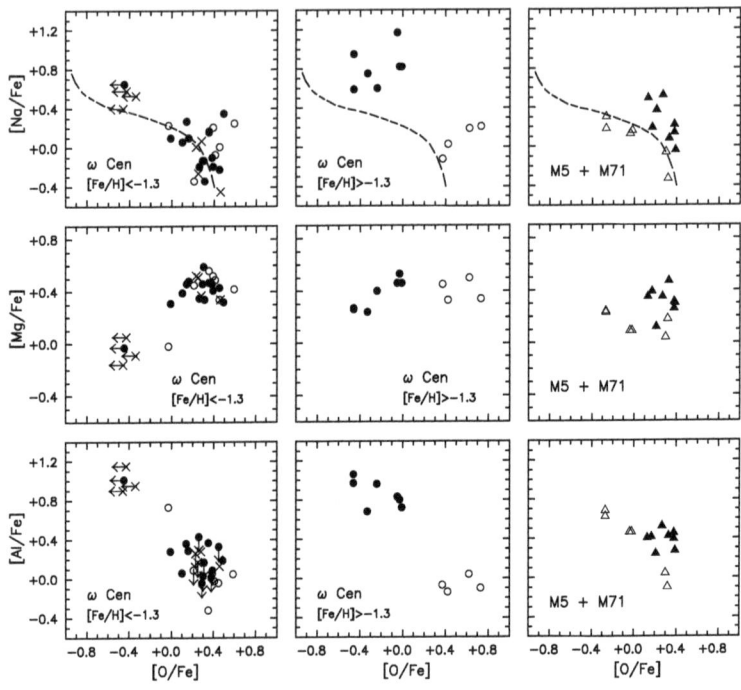

Figure 3.5. Dependences of the correlations between [Na/Fe], [Mg/Fe], [Al/Fe] and [O/Fe] on [Fe/H] for giants with [Fe/H]< −1.3 from ω Cen (left), for stars with [Fe/H]> −1.3 in ω Cen (center) and giants from the clusters M5 and M71 (right). The notation for ω Cen is the same as in Fig. 3.2. Data for Na and Al for M5 (white triangles) and M71 (black triangles) were taken from [169] and corrected by +0.05 and −0.25, respectively. The dashed curve shows the results of our calculations for Na taking into account deep mixing that can reproduce closely the observational data for M13 from Fig. 3.3a.

3.1.4 Primordial Nucleosynthesis Scenario: Inherited Anomalies

Primordial nucleosynthesis scenarios suggest that star-to-star variations in surface abundances in globular clusters did not develop in low-mass stars now observed as red giants, but instead were inherited by them from the cluster interstellar medium, which was "contaminated" by processes in stars of previous generations. Possible sources of such "contamination" include winds from high-mass MS stars, supernovae, winds from AGB stars (including the ejection of planetary nebulae) and novae. We can pose our question as follows: are there traces of nucleosynthesis that occurred in earlier generations of stars in globular clusters? If we assume that the protocluster formed from material with a cosmological chemical composition with zero metal content (i.e. the composition established immediately after the Big Bang), the answer to this question will be "definitely yes", since a broad spectrum of elements heavier than helium is observed in stars in modern globular clusters, compelling us to search for sources of primordial nucleosynthesis in clusters.

The first objects that could contribute to variations in the chemical compositions of globular clusters are massive stars, which lose mass via their stellar winds in the MS phase and later explode as type-II supernovae in the final stage of their evolution (we will not consider hypothetical supermassive objects here). However, the modern theory of stellar winds [108, 99] predicts that a massive star with $Z = 0$ should not lose a large amount of material in the process of its hydrostatic evolution [128]. Therefore, it is likely that the only substantial result of nucleosynthesis in first-generation massive stars in globular clusters is the production of elements in type-II supernovae, primarily O and other α elements [217]. We should also note here that according to [111] and [34], there is no evidence in clusters for a gradual enriching of their interstellar media in the products of the type-Ia supernovae responsible for the synthesis of iron-peak elements in models of galactic chemical evolution [199], which chronologically follow type-II supernova explosions in a cluster.

The next potential "contaminators" of the media of globular clusters are AGB stars. NDC95 reported that the abundances of heavy, neutron-enriched elements believed to be synthesised in the s-process in AGB stars with masses $1 - 3\,M_\odot$ increase with [Fe/H] in red giants in ω Cen. Since intermediate-mass stars $(3 - 8\,M_\odot)$ evolve more rapidly than objects with $M = 1 - 3\,M_\odot$, we expect that the products of nucleosynthesis in intermediate-mass AGB stars should also be present in material accreted by low-mass stars in globular clusters.

Nucleosynthesis during supernovae Apparently, the course of chemical evolution in globular clusters differs greatly from that for the chemical evolution of the Galaxy as a whole. One possible scenario for the formation of globular clusters was proposed by Cayrel [39], and was further developed by Brown et al.[25, 26]. In it, massive stars are first formed from material in the dense cluster core with a cosmological chemical composition. These stars evolve relatively rapidly (on a timescale of $\approx 10^6 - 10^7$ yrs), and explode as type-II supernovae. The supershell formed by the shock waves from multiple supernovae sweeps up and compresses the protocluster material, which is now already comprised of a mixture of matter formed in the Big Bang and during type-II supernovae. This supershell becomes a site for the birth of stars covering the entire attainable mass spectrum. There are a number of theoretical and observational arguments supporting this scenario (see the papers cited above). One of these is the *total absence* of low-mass stars with cosmological chemical compositions. We will not discuss the pros and cons of various models for the formation of globular clusters. We note only the nearly obvious fact that intermediate- and low-mass stars in globular clusters were most likely to have been formed from material that has been contaminated with the products of type-II supernovae. To determine the chemical composition of this material, we must know two things: *(i)* the mass dependence of the composition of the products supplied to the cluster interstellar medium by type-II supernovae whose progenitors were massive stars with a cosmological chemical composition, and *(ii)* an estimate of the degree of dilution of these products in the supershell.

Figure 3.6 presents the abundances of nuclides relevant to our study in material ejected by type-II supernovae as a function of the mass of the progenitor star (which can also be considered the final mass of the star, since winds from massive stars with $Z = 0$ are negligible) for metallicity $Z = 0$ according to the data of Woosley and Weaver [217]. These data were used to model the evolution of the chemical composition of the Galaxy in [199]. We should bear in mind that the chemical compositions in supernova models are less well determined for higher stellar masses. For example, the Na/O ratios obtained in models with $M = 30\,M_\odot$ can vary by a factor of 500 [217]. Such large variations in abundances with the mass

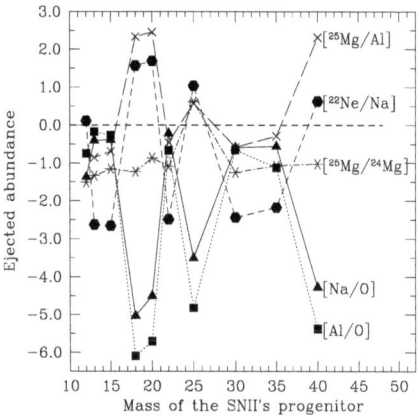

Figure 3.6. Abundances of some nuclides ejected into the interstellar medium by type-II supernovae (data from Woosely and Weaver [217]).

of the precursor are associated with uncertainties in models for the explosion, as well as the sensitivity of the precursor models to the character of the interaction between various convective zones in the star in late stages of its evolution. In spite of these large abundance variations, which can even appear random, Timmes, Woosley and Weaver [199] assert that fairly plausible results are obtained after convolution with a reasonable initial mass function and integration over time. Moreover, these results can successfully reproduce the observed evolutionary abundance variations for a large number of elements lighter than zinc in a simple model for the chemical evolution of the Galaxy.

The supernova-precursor masses in Fig. 3.6 lie in the range $12 \leq M/M_\odot \leq 40$. Following [39, 199], we consider the initial mass function in the dense protocluster core to be a Salpeter [163] mass function with a low-mass cut off, $\psi(M) \propto M^{-(1+x)}$, with a slightly lower index $x = 1.31$ (instead of 1.35), which leads to close agreement with the observed evolution of elemental abundances in the Galaxy. After

Table 3.1. Abundances of some nuclides in the interstellar medium resulting from type-II supernova explosions.

Nuclide(s)	SNeII
[C/Fe]	−0.25
[N/Fe]	−2.44
[O/Fe]	−0.05
[^{20}Ne/Fe]	+0.10
[Na/Fe]	−0.52
[^{24}Mg/Fe]	−0.05
[^{25}Mg/Fe]	−1.25
[^{26}Mg/Fe]	−1.27
[Mg/Fe]	−0.15
[Al/Fe]	−0.67
[Si/Fe]	−0.15
[Fe/H]	−2.31
[^{22}Ne/Na]	−2.34
[^{25}Mg+^{26}Mg/Al]	−0.59
^{24}Mg/^{25}Mg/^{26}Mg	98/1/1

convolving the abundances from Fig. 3.6 with $\psi(M)$ in the mass interval $12 - 40\,M_\odot$, we find the mean abundances of the elements ejected by type-II supernovae which are of interest to us. These are diluted with material with the cosmological chemical composition in the supershell. The dilution coefficient can be approximately estimated as the ratio $v_\mathrm{s}/v_\mathrm{ej} \approx 10^{-3}$ of the local sound speed in the protocluster medium $v_\mathrm{s} \approx 10$ km·s^{-1} and the initial speed of ejection of the supernova envelope $v_\mathrm{ej} \approx 10^4$ km·s^{-1} [39]. This estimate neglects the adiabatic expansion phase of the supernova envelope, and takes into account only the so-called "snowball" phase. Table 3.1 presents the abundances expected as a result of type-II supernova explosions with subsequent dilution in the supershell obtained for the scheme described above.

We can draw the following conclusions from the data in Table 3.1: *(i)* the value of [Fe/H] corresponds to those for the most metal-poor glbular clusters; *(ii)* the values of [α/Fe] are lower than those observed in globular-cluster stars – a result of the overproduction of Fe in the supernova models used (see [199]); *(iii)* the N abundance is very low; *(iv)* the ratio ^{22}Ne/Na is much lower than the initial value assumed for globular-cluster red giants (Section 3.1.3); *(v)* the abundances of ^{25}Mg and ^{26}Mg are very low relative to the abundances of ^{24}Mg and Al. An analysis of Fig. 3.6 shows that our conclusion concerning the very low ^{25}Mg/^{24}Mg ratio in the considered primordial mixture is fairly trustworthy. Thus, we are forced to search for another site of primordial synthesis of ^{22}Ne, ^{25}Mg and ^{26}Mg. Timmes, Woosley and Weaver [199] proposed that the main source of the isotopes ^{25}Mg and ^{26}Mg is intermediate-mass AGB stars, and this is supported by our calculations (see below). They also note that the abundance of N in the ejecta of type-II supernovae could be higher if dredge-up in the metal-poor massive stellar progenitors is taken into account, but suggest that AGB stars could produce the required amount of N (and C) in the Galaxy.

Products of nucleosynthesis in AGB stars To take into account the contribution of intermediate-mass AGB stars to the chemical evolution of the Galaxy, Timmes, Woosley and Weaver [199] used the results of the parametric nucleosynthesis calculations of Renzini and Voli [154], who, however, only traced variations in the abundances of ^{12}C, ^{13}C, ^{14}N and ^{16}O for $Z \geq 0.004$. Our calculations supplement the data of Renzini and Voli in two respects. First, we also considered the evolution of the abundances of elements heavier than O, and second, we adopted the lower value $Z = 10^{-4}$ and used various initial distributions of the relative abundances of nuclides within this Z, which appreciably influences the final abundances.

We took an object with mass $5\,M_\odot$ as a typical intermediate-mass AGB star. In our calculations, thermal flashes of the helium burning shell began when the mass of the carbon–oxygen core was $M_c = 0.96\,M_\odot$. We note again that we did not trace the evolution of the star on the AGB. Instead, we considered the parametric nucleosynthesis in a model star with $M = 5\,M_\odot$ in a scheme close to that used by Rezini and Voli [154] (our scheme is described in detail in [68]). The upper limit to the core mass corresponds to the Chandrasekhar limit $M_{Ch} \approx 1.4\,M_\odot$, however, it is believed that an as yet unknown instability should lead to the ejection of an envelope in the form of a planetary nebula long before M_c approaches M_{Ch}, which stops the evolution of the star on the AGB (see, for example, [153, 207]). Unfortunately, the number of flashes of the helium burning shell before the ejection of the planetary nebula is not accurately known. This number can be estimated from the observed ratio of the masses of white dwarfs and the initial masses of their progenitor stars. These ratios indicate that stars with intermediate mass and low metallicity probably spend a fairly long time on the AGB [210]. We adopted $N = 400$ as an upper limit to the number of flashes in our calculations of nucleosynthesis on the AGB. In this case, the final mass of the core is $M_c = 1.12\,M_\odot$. It is possible that this value N is an overestimate, however, our calculations show that lower values ($N = 100-200$) provide qualitatively similar (but quantitatively less pronounced) results. Table 3.1 presents the initial chemical composition used in the calculations.

Figure 3.7 depicts the final surface abundances (i.e. after 400 flashes) in an AGB star with $M = 5\,M_\odot$ for three different temperatures for the hot-bottom hydrogen burning: $T_{HBB} = 70, 90$ and $100 \cdot 10^6$ K. Note that the value of T_{HBB} is also very uncertain due to its strong dependence on depth in the hydrogen burning shell, which could reach the base of the convective envelope in the period between flashes, and this depth itself is very sensitive to the poorly known properties of the dredge-up [119]. Unfortunately, it is precisely T_{HBB} that determines whether ^{24}Mg is transformed into Al at the base of the convective envelope (Fig. 3.7).

We can conclude from Fig. 3.7 that if we neglect hot-bottom hydrogen burning (which approximately corresponds to the results depicted by the asterisks for $T_{HBB} = 70 \cdot 10^6$ K), the main new results (compared to those obtained by Renzini and Voli [154]) will be an appreciable growth in the ^{22}Ne/Na ratio and the abundances of ^{25}Mg and ^{26}Mg. The isotopes ^{22}Ne, ^{25}Mg and ^{26}Mg are primarily produced in α-capture reactions, while Na is formed in the reactions ^{22}Ne(n,γ)^{23}Ne($\beta^-\bar{\nu}$)^{23}Na. Aluminium is not synthesised during flashes, since there is no corresponding α reaction and the cross-section for neutron capture in the reaction ^{26}Mg(n,γ)^{27}Mg, which is followed by the rapid β decay ^{27}Mg($\beta^-\bar{\nu}$)^{27}Al, is very small ($\sigma_{n\gamma} \approx 0.084$ mbarn).

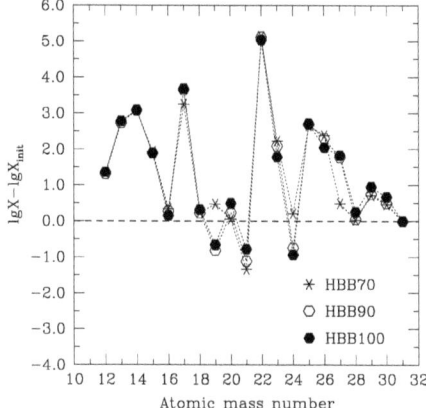

Figure 3.7. Results of nucleosynthesis of some light nuclides in intermediate-mass stars on the asymptotic giant branch after 400 thermal flashes of the helium burning shell. The abbreviation HBBT6 denotes that the hot-bottom hydrogen burning occurs at a temperature of T6·10^6 K. An atomic mass number of 26 corresponds to ^{26}Mg. The initial chemical composition was the same as in Table 3.1.

Table 3.2 compares the abundances of ^{22}Ne, Na, ^{25}Mg and ^{26}Mg established in the layer between the hydrogen and helium burning shells immediately after the 400th flash for three initial chemical compositions: (1) solar; (2) $Z = 10^{-4}$ and the distribution of relative abundances of heavy elements described in Section 3.1.3; (3) the abundances in Table 3.1. We can see that there are large increases in the ^{22}Ne/Na ratio and the ^{25}Mg and ^{26}Mg abundances for the third mixture, which was created in the framework of our model for the chemical evolution of globular clusters.

In [68], our estimate of the fraction of material initially participating in nucleosynthesis in intermediate-mass AGB stars and then captured (by accretion) by low-mass stars was $q \approx 0.1-0.2$. This estimate assumes that material that has undergone preliminary nuclear

Table 3.2. Abundances ($\lg X/X_{\text{init}}$) between the hydrogen and helium burning shells in an AGB star with mass $5\,M_\odot$ immediately after the 400th flash.

Composition	^{22}Ne	Na	^{25}Mg	^{26}Mg
Solar	2.06	1.57	1.40	1.05
$Z = 0.0001^a$	2.36	2.19	1.69	1.20
SNeII^b	4.80	1.79	2.51	1.98

[a] Relative distribution of the abundances of nuclides with the Z described in Section 3.1.3.
[b] Initial chemical composition from Table 3.1.

reprocessing in earlier generations of stars is uniformly distributed among low-mass stars. This assumption is probably valid with respect to supernovae, since turbulence behind the shock front formed as a result of multiple explosions should efficiently mix the protocluster material [26]. At the same time, the products of nucleosynthesis in AGB stars are probably distributed non-uniformly in the interstellar medium, and are preferentially captured by low-mass stars located near the AGB stars during the ejection of shells from the latter. In this case, the coefficient q could be higher.

The above discussion supports the idea that intermediate-mass AGB stars could be sources of enhanced initial abundances of ^{25}Mg (and ^{26}Mg) in globular-cluster red giants. Indeed, it follows from Fig. 3.7 that the value $q \geq 0.2$ would be sufficient to provide an increase in the ^{25}Mg abundance from $[^{25}\text{Mg/Fe}] = -0.59$ (Table 3.1) to $[^{25}\text{Mg/Fe}] > 1.0$, as required by our deep-mixing calculations (Section 3.1.3). Simultaneously, the ^{22}Ne/Na ratio would reach values near $[^{22}\text{Ne/Na}] = 0$, making it possible to synthesise Na from ^{22}Ne in quantities corresponding to the observations.

Our nucleosynthesis calculations for AGB stars with $M = 5\,M_\odot$ support the suggestion made in [199] that AGB stars can significantly enrich the interstellar medium (and, in our case, low-mass stars) in globular clusters in carbon and nitrogen. We also reiterate the conclusion drawn in [68] that intermediate-mass AGB stars may be responsible for some enrichment of low-mass stars in sodium

DEEP MIXING IN GLOBULAR-CLUSTER RED GIANTS   127

and aluminium (Fig. 3.7). However, in contrast to sodium, which is synthesised in flashes of the helium burning shell, aluminium is produced from ^{24}Mg only in hot-bottom hydrogen burning, the theoretical details of which remain uncertain. At temperatures higher than $70 \cdot 10^6$ K, the reaction ^{24}Mg(p,γ)^{25}Al occurs more rapidly than proton capture by nuclei of ^{25}Mg and ^{26}Mg. As noted above, these two isotopes are produced in large quantities during flashes. Thus, as a result of high-temperature hot-bottom hydrogen burning (for example, the evolutionary calculations of Lattanzio *et al.* [121] reach $T_{\text{HBB}} \geq 90 \cdot 10^6$ K), an excess of ^{25}Mg could be accompanied by a deficit of ^{24}Mg and increased Al abundance. However, we should bear in mind that a dilution coefficient of $q \approx 0.5$ is required to decrease the initial ^{24}Mg abundance in a low-mass star by, say, a factor of two. Of course, in this scenario, we also obtain some enhancement of the initial abundance of Al. It is important that hot-bottom hydrogen burning in intermediate-mass AGB stars *does not lead to a deficit of O* (Fig. 3.7). Consequently, deep extra-mixing is still required in red giants.

Let us summarise. The proposed primordial nucleosynthesis scenario has the following merits: *(i)* it can supply low-mass stars with a high initial ^{25}Mg abundance; *(ii)* it can explain why globular-cluster red giants with especially high Al overabundances also show ^{24}Mg deficits (because low-mass stars with low initial ^{24}Mg abundances should possess high initial ^{25}Mg abundances); *(iii)* it enriches low-mass stars in carbon (and nitrogen), as indeed observed in globular clusters (Fig. 3.2, white symbols). Clear drawbacks of the scenario are as follows: *(i)* it sometimes implies a fairly high dilution coefficient ($q \geq 0.5$), for example, to explain the low abundances of [^{24}Mg/Fe] observed in M13; *(ii)* it appears to be in contradiction with the approximate constancy of the C+N+O sum observed in some globular clusters (see Section 3.1.1), since it assumes that the initial abundances of C (and N) in low-mass stars are a function of the dilution coefficient q, which can vary from star to star; *(iii)* unfortunately, there remain many uncertainties in the scenario, which prevent us from being able to draw trustworthy conclusions.

We conclude this subsection by remarking that our calculations of nucleosynthesis via the *s*-process in AGB stars wih $M = 5\,M_\odot$ and $Z = 10^{-4}$ (which are in excellent agreement with the earlier results of Busso et al. [27]) indicate that there is no significant production of s-process elements. Consequently, we do not expect the enhanced initial abundances of ^{25}Mg in globular-cluster red giants to be correlated with excesses of s-process elements. It is currently believed that the s-process occurs in low-mass $(1-3\,M_\odot)$ AGB stars, in which the neutrons required for the s-process are supplied by the reaction ^{13}C$(\alpha,n)^{16}$O, and not the reaction ^{22}Ne$(\alpha,n)^{25}$Mg, as in our case (see Gallino et al. [80]). Therefore, the observational data of NDC95 showing the abundances of some s-process elements (Y, Ba, La and Nd) in giants in ω Cen to increase with [Fe/H] was interpreted in that paper as evidence for a primordial enrichment of low-mass stars by material ejected by AGB stars with masses $1-3\,M_\odot$. However, in this case, where are the signs of intermediate-mass AGB stars, which evolve much more rapidly? Our scenario provides an answer to this question.

Solution using a "black box" model If the Al excesses observed in red giants with deficits of [Mg/Fe] in M13 and ω Cen have indeed arisen from ^{24}Mg and not ^{25}Mg, and if nuclear physicists confirm[7] the correctness of the rates of the reaction ^{24}Mg$(p,\gamma)^{25}$Al from [36], the only possible explanation for the anticorrelation between [Mg/Fe] and [Al/Fe] is hydrogen burning at a temperature that is much higher than that of the hydrogen burning shell in standard models ($T_6 \leq 55$).

To test this idea, we considered hydrogen burning at constant temperature and density (a "black box" model). We adopted $\rho = 44.7$ g·cm^{-3} for the density, as in the deep-mixing studies of Langer et al.[115]. The initial chemical composition was taken to be that described in Section 3.1.3. The nucleosynthesis calculations were stopped when 5% of the hydrogen was burned. Furthermore, the calculated abundances were mixed with material with the initial chemical composition, in fractions from 0 to 100%. The temperature was treated as a free parameter. The results are shown in Fig. 3.8, where the curves represent the theoretical dependences between the abundances of O, Na, Mg and Al in the mixture considered. The crosses on the solid curves correspond to mixtures in which the fraction of the

---

[7] Such confirmation is provided by the new (NACRE) data on reaction rates for the MgAl cycle [3].

DEEP MIXING IN GLOBULAR-CLUSTER RED GIANTS 129

initial material decreases from $q = 1$ to 0.1 in steps of 0.1 (from right to left). The range of temperatures yielding the observed anticorrelation of [O/Fe] with [Na/Fe] and [Al/Fe] and correlation of [O/Fe] with [Mg/Fe] is strikingly narrow: $T_6 = 70$ (short-dashed curve), $T_6 = 74$ (solid curve) and $T_6 = 78$ (dot–short-dashed curve). Thus, if this idea were correct, we could estimate very precisely the temperature for hydrogen burning that would give rise to the observed relations between the O, Na, Mg and Al abundances in globular-cluster red giants: $T_6 = 74 \pm 2$! Our choice of the density does not affect this estimate, and if the calculations are continued until more than 5% of the hydrogen is burned, it becomes simply impossible to reproduce all three correlations with a single temperature. It is interesting that a very similar result (with a slightly higher temperature $T_6 = 78$) is obtained if we consider hydrogen burning in a high-mass convective core (we modelled the structure of such a core using a polytrope with index $n = 1.5$, with the temperature in the centre of the core being a free parameter).

The next question we must answer is what type of stars could be identified with the "black box" model described above? We considered a model for a zero-age main-sequence star with mass 125 $M_\odot$ and $Z = 4 \cdot 10^{-4}$, but found that its central temperature was $T_6 = 53$, which is too low. In general, any conceivable "black box" model corresponding to objects from earlier generations in globular clusters encounters difficulties in explaining why material was ejected into the interstellar medium after only 5% of the hydrogen was burned. Another problem is understanding how some low-mass stars could capture up to 90% of the material ejected by the "black box" (Fig. 3.8). However, these two problems can be easily solved if we place the "black box" inside a star ascending the red giant branch. Of course, in this case, we must understand the mechanism that increases the temperature of the hydrogen burning shell to $T_6 = 74$. Langer et al. [116] proposed that one possible mechanism could be thermal instability in the hydrogen burning shell (see also [79]).

However, a high-temperature origin for the anticorrelation between [Mg/Fe] and [Al/Fe] in globular-cluster red giants is not consistent with the analysis of the isotopic composition of Mg in M13 carried out by Shetrone [170], since not only ^{24}Mg, but also ^{25}Mg and ^{26}Mg are rapidly destroyed at $T_6 = 74$. For example, if we begin

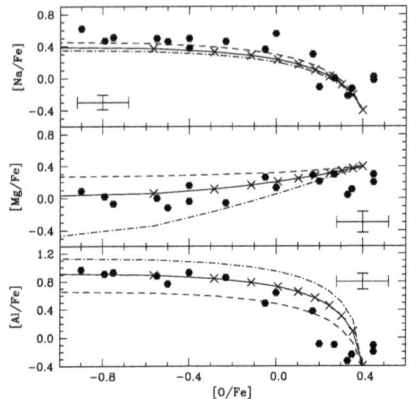

Figure 3.8. Relationships between the abundances of O, Na, Mg and Al (curves) in mixtures with fractions of material that has not undergone nuclear reprocessing varying from 0% to 100% (the crosses correspond to 100%, 90%, 80%, ..., from right to left). In the material that has undergone nuclear reprocessing, we considered hydrogen burning at constant density ($\rho = 44.7$ g·cm^{-3}) and temperature, and continued the calculations until 5% of the initial hydrogen content was burned. The temperature ($T_6 = 70$ – dashed curves, $T_6 = 74$ – solid curves, $T_6 = 78$ – dot–dashed curves) was selected in order to reproduce the corresponding correlations in giants in M13 (hexagons). The initial chemical composition was that described in Section 3.1.3.

with a total abundance [(^{25}Mg+^{26}Mg)/Fe] = 0 and the isotopic ratios ^{24}Mg/^{25}Mg/^{26}Mg = 90/4.5/5.0 (as in the mixture described in Section 3.1.3, and which is very close to the ratios observed by Shetrone [170] in the "unmixed" giant L 598 in M13), we obtain after burning 5% of the hydrogen [(^{25}Mg+^{26}Mg)/Fe] = −0.50 (while Shetrone gives +0.21) and ^{24}Mg/^{25}Mg/^{26}Mg = 92.3/4.7/3.0 (compare with the mean ratios of Shetrone, 56/22/22).

3.1.5 Conclusion

We have shown that deep mixing in stars ascending the red giant branch can provide a self-consistent explanation for the anomalous abundances of C (as well as the ratio $^{12}C/^{13}C$), N, O and Na in giants in ω Cen and the global anticorrelation between [O/Fe] and [Na/Fe], which is most clearly expressed by giants in M13. There are other observational manifestations of extra-mixing in globular-cluster red giants, the most direct being the gradual decrease in [C/Fe] with increasing luminosity on the red giant branch observed for a number of globular clusters (see corresponding references in Section 3.1.1). Similar evidence was discovered by Pilachowski *et al.* [146]: the Na abundances in the M13 stars they considered become higher with approach to the tip of the red giant branch.

In this section, we have not discussed the mechanism for extra mixing in red giants. The most promising candidate mechanism is probably some kind of instability induced by rotation (such as a barocline or shear instability; see Section 2.2). In this case, clusters with especially strong star-to-star variations in surface abundances should have the most rapidly rotating stars. Indeed, the horizontal branch stars in M13 show unusually rapid rotation for their class [145]. The cluster ω Cen possesses one of the most blue horizontal branches among known globular clusters [216], which may provide evidence for surface overabundances of He in the stars on this branch, which they acquired on the red giant branch as a consequence of deep mixing [159, 191].

Unfortunately, scenarios with deep mixing alone cannot explain the anticorrelations between the Mg and Al abundances in giants in M13 and ω Cen; this requires some additional assumptions. Possible solutions include a strong, but not yet detected, low-energy resonance in the reaction $^{24}Mg(p,\gamma)^{25}Al$ and episodic growth in the temperature in the hydrogen burning shell to values $T_6 \approx 74$ (in standard models, $T_6 \leq 55$) in stars ascending the red giant branch. The first possibility is likely to call forth objections from nuclear physicists, while the second is in contradiction with an analysis of the magnesium isotopic composition for a small sample of bright M13 giants [170] (see Section 3.1.4).

On the other hand, there are convincing observational arguments in support of primordial nucleosynthesis scenarios. To those discussed in Section 3.1.4 we can add the well-known bimodality in the distribution of the CN molecular-band intensities in stars in 47 Tuc, which can be traced right to the main-sequence turnoff [19]. Moreover, in 1996, Briley *et al.* [20] showed that strong CN bands (and weak CH bands) are accompanied by spectral indications of enhanced Na abundances in stars located in the immediate vicinity of the main sequence in 47 Tuc. In our work, we have attempted to determine whether extra-mixing in low-mass MS stars can lead to underabundances of C and overabundances of N and Na without carrying too much He to the stellar atmosphere (which would lead to a conflict with the very narrow colour–magnitude diagram near the MS turnoff in 47 Tuc [203]). It turned out that it is not possible to satisfy these conditions, even using the new rates for the NeNa-cycle reactions of El Eid and Champagne [72]. Consequently, the bimodality in the CN-band intensity distribution and the Na excesses in stars in 47 Tuc probably have a primordial origin, possibly associated with AGB stars.

We have proposed a combined scenario: anomalies in chemical composition inherited by low-mass stars plus deep extra-mixing in these stars. In this scenario, intermediate-mass AGB stars are considered to be the source providing the ^{25}Mg excesses and ^{24}Mg deficits in globular-cluster red giants. In this case, the anticorrelation between [O/Fe] and [Al/Fe] in giants in M13 and ω Cen can be reproduced fairly well in calculations with extra-mixing. In this picture, Al is synthesised from ^{25}Mg (for a modification of this scenario, see Section 3.2).

3.2 Contribution of ^{26}Al to the Anticorrelation Between the O and Al Abundances in Globular-Cluster Red Giants (2001)

3.2.1 Introduction

In [66] (see also Section 3.1), we proposed a combined scenario that can explain the star-to-star variations in the abundances of C, N, O, Na and Al in globular-cluster red giants in a self-consistent way. This scenario includes two components: deep mixing (or an evolutionary component) and primordial, or inherited, anomalies in chemical composition. The deep-mixing component shows that overabundances of N, Na and Al correlated with underabundances of C and O can, in principle, arise in evolving red giants as a result of hydrogen shell burning, via the competing reactions of the CNO, NeNa and MgAl cycles. However, this requires that some extra-mixing connects the outer part of the hydrogen burning shell and the base of the convective zone. For this to give rise to large (>1 dex) variations in [Al/Fe] and for the anticorrelation between O and Al to resemble that observed for red giants in ω Cen [139], we must suppose in addition that: *(i)* the initial abundance of ^{25}Mg in these stars was enhanced to [^{25}Mg/Fe] $=1.1$ and *(ii)* the rate of the reaction ^{26}Al^g(p,γ)^{27}Si, which opens a channel for the production of ^{27}Al via the rapid decay ^{27}Si($\beta^+\nu$)^{27}Al, is actually a factor of 10^3 higher than the value recommended by Caughlan and Fowler in 1988 [36]. The first assumption is supported by our adopted working model for the chemical evolution of globular clusters. In this model (based on the earlier studies [39] and [26]), the type-II supernovae in a globular cluster enrich the interstellar medium in iron and α elements (such as ^{16}O and ^{24}Mg), while intermediate-mass stars ($M = $ 3-8 $M_\odot$) on the asymptotic giant branch enrich the medium in the isotopes ^{25,26}Mg and, to some extent, in Na and Al. The second assumption was admissible since the uncertainty in the rate of the reaction ^{26}Al^g(p,γ)^{27}Si at that time reached a factor of $\sim 10^3$ (see [5]).

The following important results related to the origin of the O–Na and O–Al anticorrelations in globular-cluster red giants have been obtained since we wrote [66]. *(i)* The model we used to describe the chemical evolution of globular clusters has received new observational support [98, 181] and further theoretical development [143]. *(ii)* Detailed calculations of nucleosynthesis in intermediate-mass, low-metallicity AGB stars have confirmed our earlier conclusion that significant amounts of ^{25}Mg and ^{26}Mg may be produced in them [120, 134]. *(iii)* New data on nuclear reaction rates in stars (the NACRE data) have been published [3]; the upper limit for the rate of the reaction ^{26}Al^g(p,γ)^{27}Si allowed by these data is only a factor of ~ 50 higher than the value given by [36] for temperatures $T = 40$–$50\cdot 10^6$ K, characteristic of the outer regions of the hydrogen burning shell in globular-cluster red giants. *(iv)* Until recently, ω Cen remained the only globular cluster in which four correlations between surface abundances (O–Na, O–Al, C–O and C–N) were directly observed for a group of red giants, but this is now also true of the cluster M4 [96]. *(v)* Recently, the O–Na anticorrelation, and possibly also the O–Al anticorrelation, both of which were observed earlier only in globular-cluster red giants, has been detected in stars below the MS turnoff point in the cluster NGC 6752 [86]. *(vi)* Ventura et al. [205] interpreted the presence of an O–Na anticorrelation for stars near the main sequence using the model for the chemical evolution of globular clusters referred to above and their own discovery that, at very low metallicities, hot-bottom burning (HBB) in intermediate-mass AGB stars occurs at a temperature $T_{\rm HBB} \geq 10^8$ K. Ventura and colleagues suggest that O should burn at such high temperatures, which should also increase the Na abundance, so that the O–Na anticorrelation in globular-cluster stars is mostly likely to be a consequence of the accretion by MS dwarfs of matter lost by intermediate-mass AGB stars of earlier generations (i.e. they suggest that only the second component of the combined scenario operates).

In this section, we modify the combined scenario to bring it into agreement with these new experimental and observational results.

3.2.2 The Problem of the O Deficit

The presence of heavy elements synthesised in the *s*-process in the atmospheres of red giants in ω Cen and M4 [181, 96] undoubtedly indicates that AGB stars have contributed to the production of anomalies in the chemical composition of globular-cluster red giants. Other obvious evidence for the important role of primary nucleosynthesis and the subsequent accretion of the products of this nucleosynthesis by low-mass stars in globular clusters is provided by variations in the C, N, O and Na abundances of stars below the MS turnoff point [20, 86]. On the other hand, the decrease in [C/Fe] with decreasing $M_{\rm V}$ in red giants in M92, first detected by Langer et al. in 1986 [117] and recently confirmed in [10], provides convincing proof of the action of extra-mixing in globular-cluster red giants. Moreover, Gratton et al. [87] have shown that extra-mixing leading to appreciable variations in the surface abundances of C, N and Li in metal-poor field red giants is probably the rule rather than the exception. Therefore, the main questions we must address are the following: *(i)* How deep is "deep mixing" in globular-cluster red giants, or in other words, is this mixing able to penetrate deep enough into the hydrogen burning shell to carry out material depleted in O and enriched in Na and Al? *(ii)* Of elements showing anomalous abundances in globular-cluster red giants, which could be produced or destroyed in appreciable quantities in primordial nucleosynthesis (i.e. in stars of earlier generations) in globular clusters, and in which types of stars?

The most complex problem in any scenario taking into account only inherited anomalies in chemical composition is to explain the very low O abundances observed in some globular-cluster red giants. For example, some red giants in M13 have [O/Fe] ≈ -0.8 [107]. In scenarios incorporating only primordial nucleosynthesis, this would mean that: *(i)* being MS dwarfs in the past, such stars should have [O/Fe] ≤ -0.8; *(ii)* the low O abundances cannot be purely surface anomalies, and almost the entire volume of such a MS dwarf must have [O/Fe] ≈ -0.8. Indeed, the mass of the convective envelope of a star with $M = 0.8\,M_\odot$ (a typical mass for a globular-cluster red giant) with relative abundances of hydrogen and heavy elements $X = 0.24$ and $Z = 0.005$ does not exceed $0.03\,M_\odot$ on the main sequence, and reaches $0.53\,M_\odot$ on the red giant branch. If we assume that a MS dwarf has [O/Fe] $= -0.8$ only in its convective envelope (for example, as a consequence of accreting surface layers of material with a low O content), while its remaining volume has [O/Fe] $= +0.4$ (characteristic for both the majority of globular-cluster red giants and Population II field dwarfs), subsequent dilution in the convective envelope of the red giant leads to the value [O/Fe] $=+0.38$, contrary to observations. Thus, pure primordial nucleosynthesis scenarios would require a MS dwarf with initial mass $M = 0.3\,M_\odot$ and [O/Fe] $= +0.4$ to accrete up to $0.5\,M_\odot$ of material with [O/Fe] < -0.8.

Such large amounts of accreted material are possible from a theoretical point of view in the dense cores of globular clusters [198]. However, in accretion scenarios, we must also identify a source of material in the interstellar medium that simultaneously possesses fairly large deficits of O and excesses of Na and Al. Ventura *et al.*[205] proposed that this source could be intermediate-mass AGB stars that lose mass via their stellar winds. At the very high temperatures at the base of the convective envelope of an intermediate-mass, low-metallicity AGB star obtained in their evolutionary calculations ($T_{\text{HBB}} \geq 10^8$ K), the

O abundance will be significantly decreased throughout the envelope. Ventura et al. [205] *assumed* (without performing the corresponding nucleosynthesis calculations) that an excess of Na will form in these stars simultaneously with the decrease in their O abundances. However, apparently this should not happen.

Our Fig. 3.9 shows the final abundances for isotopes of the CNO, NeNa and MgAl cycles in the convective envelope of a model AGB star with $M = 5\,M_\odot$ after 20 thermal flashes of the helium burning shell. These abundances were obtained using our code for parametric nucleosynthesis calculations in AGB stars [68], but using the new NACRE data for the nuclear reaction rates. The initial abundances ($X_{\text{init}}(A)$) were taken to be those obtained as a result of multiple type-II supernova explosions in the protocluster cloud (Table 1 from [66]). The abundance variations in Fig. 3.9 were calculated for $T_{\text{HBB}} = 70 \cdot 10^6$ K (x's), $T_{\text{HBB}} = 85 \cdot 10^6$ K (white squares) and $T_{\text{HBB}} = 10^8$ K (black squares). We can see that at temperatures T_{HBB} for which the abundance of O ($A = 16$) decreases, the abundance of Na ($A = 23$) either decreases or only very slightly grows. This behaviour for Na is due to its destruction in the reaction ^{23}Na(p,α)^{20}Ne, which becomes efficient at temperatures $T \geq 90 \cdot 10^6$ K. Thus, it is unlikely that intermediate-mass AGB stars supply the interstellar media of globular clusters with material with anticorrelated O and Na abundances. At the same time, the [^{22}Ne/Na], [^{25,26}Mg/^{24}Mg] and [^{25,26}Mg/^{27}Al] ratios increase appreciably (at $T_{\text{HBB}} > 70 \cdot 10^6$ K, Fig. 3.9) – a result that is an important ingredient of our combined scenario (see [66]).

To the above we can add two arguments against the hypothesis of Ventura et al. [205]. First, as we can see in Fig. 3.9, the ^{24}Mg abundance should decrease in proportion to the ^{16}O abundance, or even slightly more rapidly. In pure primordial nucleosynthesis scenarios, this should imply that *all* globular-cluster red giants with O deficits should also be underabundant in ^{24}Mg (this is usually the most abundant isotope of magnesium). However, at least in red giants in M4, the observed [Mg/Fe] values are as high (their mean is +0.44) as the mean abundances of α elements in field and halo dwarfs, and there is a total absence of a correlation between [O/Fe] and [Mg/Fe]. Second, Cohen [48] did not detect variations in the intensities of CN and CH molecular bands in stars below the MS turnoff in M13, which

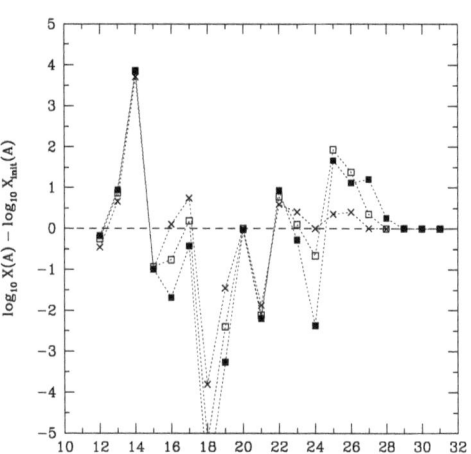

Figure 3.9. Final and initial abundances compared for CNO, NeNa and MgAl isotopes in the convective envelope of a model AGB star with mass $5\,M_\odot$ after 20 flashes of the helium burning shell. Calculations were carried out for temperatures at the base of the envelope $T_{\mathrm{HBB}} = 70 \cdot 10^6$ K (x's), $T_{\mathrm{HBB}} = 85 \cdot 10^6$ K (white squares) and $T_{\mathrm{HBB}} = 10^8$ K (black squares).

probably excludes the presence of inherited anomalies in the C and N abundances in this cluster, in spite of the fact that some red giants in M13 have very low values of [O/Fe] that are anticorrelated with both [Na/Fe] and [Al/Fe].

We propose an alternative explanation for the O–Na and O–Al anticorrelations in stars near the main sequence in NGC 6752, which not only does not rule out the hypothesis of deep extra-mixing in globular-cluster red giants, but actually invokes such mixing. Namely, we suggest that some MS dwarfs in globular clusters have accreted material lost by red giants that are the primary components of close binary systems in the so-called common envelope stage of their evolution (see [94]). If these red giants underwent extra-mixing before they filled their Roche lobes, the envelope material lost by them could be enriched in Na and Al and underabundant in O. Potentially any primary component of a binary with initial mass $0.8 < M/M_\odot < 2.5$ could play a role in this scenario, since the hydrogen burning shell in stars with such masses smooths out the jump in the molecular-weight distribution formed by the base of the convective envelope (which is thought to hinder the action of extra-mixing) before these stars reach the tip of the red giant branch (see, for example, [42, 45]).

3.2.3 The Problem of the Al Excess

The new (NACRE) rate of the reaction $^{26}\text{Al}^\text{g}(\text{p},\gamma)^{27}\text{Si}$ (even an upper limit) is not sufficiently high to produce ^{27}Al (which is a product of the β^+ decay of ^{27}Si) in globular-cluster red giants. Nearly all the initial ^{25}Mg, whose initial abundance is assumed to increase during primordial nucleosynthesis (in previous generation intermediate-mass AGB stars), is transformed into $^{26}\text{Al}^\text{g}$, which, in turn, is more likely to decay into ^{26}Mg than to capture a proton to form ^{27}Al. Therefore, the NACRE data imply that the only possible explanation for the O–Al anticorrelation in globular-cluster red giants is that the observed [Al/Fe] variations are in fact due to star-to-star variations in the surface abundance of the radioactive isotope $^{26}\text{Al}^\text{g}$ produced by deep mixing (Fig. 3.10b and 3.11b). The following simple estimates support this hypothesis. The linear size of the radiative zone between the hydrogen burning shell and the base of the convective envelope in globular-cluster red giants is $\Delta r = 1 - 2\,R_\odot$. With a diffusion coefficient $D_\text{mix} = 4 - 5 \cdot 10^8$ cm$^2 \cdot$s^{-1}, which can simultaneously reproduce all four correlations between the surface abundances of C, N, O, Na and Al in red giants in ω Cen and M4, the characteristic extra-mixing time is $\tau_\text{mix} \approx (\Delta r)^2/D_\text{mix} = 0.3 - 1.5 \cdot 10^6$ yrs. This is comparable to the life span of $^{26}\text{Al}^\text{g}$ ($1.07 \cdot 10^6$ yrs). Consequently, some freshly synthesised $^{26}\text{Al}^\text{g}$ could easily "survive" transport from the hydrogen burning shell to the base of the convective envelope. Our detailed nucleosynthesis calculations taking into account extra-mixing (the corresponding code is described in [64]) confirm these estimates (Fig. 3.10b and 3.11b). Note that, in spite of the fairly prolonged life time of the stars on the red giant branch ($\sim 10^7$ yrs for globular-cluster red giants), their surface abundances of $^{26}\text{Al}^\text{g}$ continue to monotonically increase (their absolute values

DEEP MIXING IN GLOBULAR-CLUSTER RED GIANTS 139

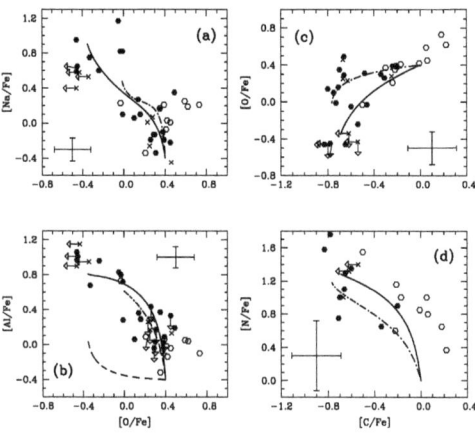

Figure 3.10. Correlations between the surface abundances of C, N, O, Na and Al in red giants in ω Cen (the meaning of the symbols is explained in Section 3.1; the observational data are from [139]) together with the results of our deep-mixing calculations for two sets of values for the mixing depth (see Section 2.1 for a definition of $\delta m_{\mathrm{mix}} \equiv \delta M_{\mathrm{mix}}$) and rate (diffusion coefficient in units of cm$^2 \cdot$s^{-1}), (δm_{mix}; D_{mix}): $(0.05; 5 \cdot 10^8)$ – solid and dashed curves (in panel **b**), $(0.065; 5 \cdot 10^8)$ – dot–dashed curves. The initial ^{25}Mg abundance was taken to be [^{25}Mg/Fe] $= 1.2$. In panel **b**, the solid and dot–dashed curves show the production only of the isotope ^{26}Al^{g}, while the dashed curve shows that only of ^{27}Al. The large crosses indicate the characteristic observational errors.

reach $X(^{26}\mathrm{Al}) = 9.4 \cdot 10^{-6}$ in Fig. 3.10b, solid curve) since there is an operating source of ^{26}Al^{g} at the base of the region of extra-mixing throughout this time. There is also some increase in the ^{27}Al abundance, but only much later (Fig. 3.10b, dashed curve).

3.2.4 Concluding Remarks

Thus, the new NACRE data on nuclear reaction rates in stars [3] have compelled us to modify our combined scenario [66]. We have

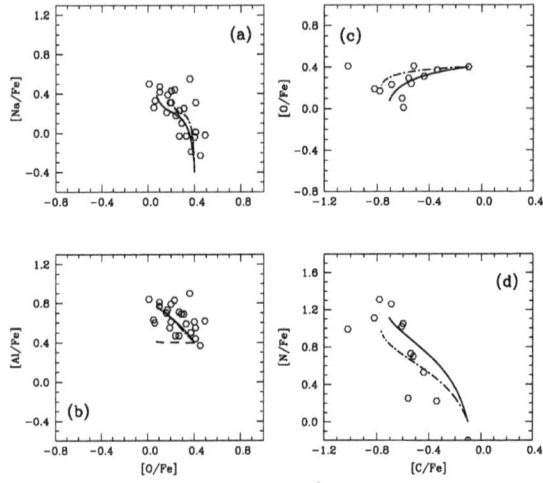

Figure 3.11. Correlation between the surface abundances of C, N, O, Na and Al in red giants in M4 (observational data from [96]) together with results of our extra-mixing calculations for two sets of values for the mixing depth and rate (δm_{mix}; D_{mix}): $(0.065; 4 \cdot 10^8)$ – solid and dashed (in panel **b**) curves, $(0.075; 4 \cdot 10^8)$ – dot–dashed curves. The initial abundances of ^{12}C, ^{25}Mg and ^{27}Al were taken to be [C/Fe] $= -0.1$, [^{25}Mg/Fe] $= 1.2$ and [Al/Fe] $= +0.4$. In panel **b**, the solid curve shows the production of only the isotope ^{26}Al^{g}, while the dot–dashed curve shows that for only ^{27}Al. The nucleosynthesis calculations for this figure and Fig. 3.10 were carried out using the new NACRE data for the nuclear reaction rates [3].

re-examined only that component of the scenario concerned with deep mixing. We have established that, when the NACRE reaction rates are used, *all* the (anti)correlations between elemental abundances observed in giants in ω Cen and M4 can be explained in a natural way in a combined scenario if the observed Al excess is, in fact, associated with the radioactive isotope ^{26}Al^{g}. Its presence at the surface of a red giant could be maintained by extra-mixing that reaches layers of the hydrogen burning shell in which this isotope is produced. As before, the component of the scenario associated with primordial nucleosynthesis is responsible for the enhanced initial abundance of ^{25}Mg (believed to be produced and ejected into the interstellar medium of the cluster by intermediate-mass AGB stars), from which ^{26}Al$^{\rm g}$ is synthesised.

It is important that when the temperature at the base of the convective envelope exceeds $8 \cdot 10^7$ K, the accumulation of ^{25}Mg in intermediate-mass, low-metallicity AGB stars is accompanied by the production of N and ^{27}Al and the destruction of ^{16}O, Na (in the reaction ^{23}Na(p,α)^{20}Ne) and ^{24}Mg. In particular, a decrease in the ^{16}O abundance probably rules out the possibility of an accompanying increase in the Na abundance (see Fig. 3.9). At lower temperatures, only N, Na and ^{25}Mg (among nuclides related to abundance anomalies in globular clusters) are produced in appreciable quantities. Consequently, pure primordial nucleosynthesis scenarios cannot explain the O–Na anticorrelation, especially in the region where [O/Fe] $\leq$ −0.4 and [Na/Fe] $\geq$ +0.4 (see, for example, Fig. 3.10a). Therefore, we associated the presence of this anticorrelation in globular-cluster red giants with deep mixing on the red giant branch (Fig. 3.10a and 3.11a), and in stars below the MS turnoff and in sub-giants with the accretion by MS dwarfs of material that was involved in extra-mixing in more massive ($0.9 \leq M/M_\odot \leq 2.5$) red giants and then lost by them, for example, in the common-envelope stage of evolution of binary systems. We consider it improbable that ^{26}Al$^{\rm g}$ associated with deep mixing on the red giant branch was preserved in material accreted by the MS dwarfs, since this would imply that the corresponding mass loss and subsequent accretion occurred no more than $\sim 10^6$ yrs ago (otherwise, all the ^{26}Al$^{\rm g}$ would long ago have decayed into ^{26}Mg). Therefore, the large excesses of Al ([Al/Fe] $>$ +0.6) observed for two (out of nine) sub-giants in NGC 6752 are most likely the result of contamination of these stars with material enriched in ^{27}Al that was lost by the most massive intermediate-mass AGB stars. This hypothesis is supported by the fact that only (and precisely) these two sub-giants have negative [Mg/Fe] ratios (i.e. they probably have underabundant ^{24}Mg, see [66]).

However, as many as five of the remaining seven sub-giants in NGC 6752 possessing moderate excesses of Al and positive [Mg/Fe] values could have fairly high ^{25}Mg abundances, since ^{25}Mg is produced in all (not only the most massive) AGB stars with intermediate mass and low metallicity. This initial ^{25}Mg excess can further be available for the production of ^{26}Al^{g} on the red giant branch. In turn, freshly synthesised ^{26}Al^{g} can be carried to the surface of the giant by extra-mixing, whose action is characterised by a deficit of O that is correlated with an excess of Al. In connection with this, we note that all nine of the sub-giants in NGC 6752 have either excess or undetermined O abundances.

Our various proposed interpretations for the overabundances of Al in globular-cluster red giants and stars below the MS turnoff are indirectly supported by the following observations.

(i) There are two groups of red giants in ω Cen. In both groups, the giants demonstrate a well-defined O–Al anticorrelation (they have [O/Fe] < 0 and [Al/Fe] > +0.4), but the giants in the first group have [Mg/Fe] $\leq$ 0, while those in the second group have [Mg/Fe] $\geq$ +0.3 (see Fig. 3.5 in Section 3.1). It is quite possible that the stars in the first group are already initially enriched in Al (in the form of ^{27}Al), similar to the two NGC 6752 sub-giants possessing the most clearly expressed Mg–Al anticorrelation, while the stars in the second group acquired their Al excesses only on the red giant branch, as a result of the production of ^{26}Al^{g} from ^{25}Mg in extra-mixing.

(ii) It is known that star-to-star variations in C and N abundances in globular clusters can be present in stars near the main sequence, as, for example, in NGC 6752 [189]. This could be associated purely with primordial nucleosynthesis. On the other hand, the correlation between [C/Fe] and M_V observed in M92 (and a number of other clusters) [10] undoubtedly has an evolutionary origin. Consequently, the manifestations of extra-mixing and primordial nucleosynthesis may imitate each other (similar to the overabundances of ^{27}Al and ^{26}Al^{g} discussed above). The reason is that both components of the combined scenario are associated with hydrostatic hydrogen burning.

There are two main possibilities for testing our interpretation of the origin of Al-abundance anomalies in globular-cluster red giants: estimation of the ^{27}Al/^{26}Al^{g} isotopic ratio and detection of the γ-ray line (with energy 1.8 Mev) formed during the decay of ^{26}Al^{g}. Simple estimates show that the expected flux of γ rays from ω Cen is of order 10^{-7} photons/(cm^2·s). This flux is two orders of magnitude below the detection threshold of the COMPTEL space γ-ray telescope, but may be detectable using instruments from the next generation of γ-ray telescopes.

4. Transition of MS Stars with Masses of 10 and 30 $M_\odot$ to States with Stationary Rotation (1998)

4.1 Introduction

Massive ($M \geq 10\,M_\odot$) main-sequence stars, in whose convective cores H is transformed into He, are observed as bright O and B stars. Over the past two decades, much observational data indicating the action of extra-mixing in these stars has been accumulated (see, for example, the review by Lyubimkov [126]). The most convincing arguments supporting this conclusion are the excesses of N and He observed in the atmospheres of OB stars. Lyubimkov [123, 124] was the first to discover overabundances of N (and even He) in the atmospheres of most single early B-type MS stars, and to show that they were correlated with the age and mass of the star. The subsequent verification of these results by Gies and Lambert [83] confirmed Lyubimkov's main conclusions concerning N.

Substantial He excesses (to three times the solar He abundances) were also detected in bright OB stars (most of which are still located on the main sequence) by the Munich group [91]. These are, on average, more massive than the early B stars studied by Lyubimkov. It turned out that the anomalously high He abundances in OB stars are accompanied by a discrepancy in estimates of their masses acquired using two different methods. Namely, the "evolutionary masses" obtained by comparing the stars' positions on the Hertzsprung–Russell diagram with theoretical evolutionary tracks were systematically higher than the "spectroscopic masses" derived from their spectra and independently from stellar-wind theory.

Langer [113], Weiss [211] and Denissenkov [59] proposed that both the anomalous helium abundances and the discrepancy in the mass estimates could be due to the effect of extra-mixing (of unknown origin) in the radiative envelopes of OB stars. Stellar material enriched in helium becomes more transparent to radiation, leading to an increase in the star's luminosity ($L \propto \mu^4$, where μ is the mean molecular weight [102]). As a result, the evolutionary mass will be overestimated. Note that, in their subsequent publication, Herrero [90] reduced the mass discrepancy he and his coauthors had published earlier [91]; however, the old values for the He excesses remained unchanged.

A massive MS star has a large convective core surrounded by a radiative envelope. Recall that He is essentially not produced outside the core (see Fig. 4.3a below). Therefore, the mechanism for the extra-mixing must enable the mixing to penetrate to the core and allow a mixing rate sufficiently high for matter in the deep interior to be able to reach the surface layers over the life span of the OB star on the main sequence.

It is known that MS OB stars rotate rapidly, and it is natural to associate extra-mixing in their radiative envelopes with this rotation. In 1992, Zahn [219] published a new scheme describing how extra-mixing can develop and be maintained in the radiative zone of a single non-magnetic rotating star. Zahn's only assumption was that the turbulence arising due to various instabilities associated with the differential rotation of the star was strongly anisotropic, so that the horizontal component of the corresponding coefficient of turbulent diffusion was much greater than the vertical component. Among various instabilities associated with rotation, Zahn chose shear instability as having one of the shortest growth times. It was assumed that shear flows come about due to meridional circulation [71, 206, 190] as a consequence of the redistribution of angular momentum by this circulation.

The main assumption of the dominance of the horizontal turbulence makes it possible to treat the star as a collection of layers rotating with angular velocity Ω that depends only on the distance r from the centre (differences in the rotational velocity within an arbitrarily chosen spherical surface will be suppressed by the high horizontal turbulent viscosity). Zahn derived an equation describing the redistribution of angular momentum under the joint action of meridional circulation and (vertical) turbulent diffusion (which is also thought to arise due to shear instability). The solution of this equation can be used to trace the evolution of the $\Omega(r)$ profile in the rotating star. The presence of stronger horizontal turbulence must also be taken into account when considering the transport of elements by meridional circulation. Chaboyer and Zahn [40] showed

DEEP MIXING IN GLOBULAR-CLUSTER RED GIANTS

that horizontal turbulence plays a role in the erosion in this process, leading to an exchange of chemical elements between rising and sinking meridional-circulation flows, which strongly reduces its efficiency in radial mixing.

In subsequent years, several important revisions and additions were introduced into Zahn's scheme: *(i)* the contribution of the shear instability in horizontal planes (spherical surfaces) to turbulent diffusion in the vertical direction was no longer included, since shear flows in these planes should be efficiently suppressed by the large horizontal turbulent viscosity [220]; *(ii)* a new expression for the vertical component of the turbulent viscosity coefficient was obtained in [195], taking into account the influence of horizontal erosion and energy loss by turbulent elements on the emitted radiation; *(iii)* in 1998, Maeder and Zahn [132] proposed a modification of Zahn's original scheme, intended to take into account the evolution of the star (i.e. time variations in the distributions of T, ρ etc.) and the joint action of meridional circulation and semi-convection.

Zahn's scheme provides a self-consistent solution to the problem of extra-mixing in the radiative zones of rotating stars, in the sense that the associated rotation profile $\Omega(r)$, which determines the diffusion coefficient (see below), is no longer chosen arbitrarily (for example, $\Omega(r) = \mathrm{const}$, as was often assumed in earlier studies), and is instead obtained by solving an equation describing the redistribution of angular momentum in the star due to the same mixing. A similar algorithm was used in [73, 148, 75]; however, in these studies, the transport of angular momentum by meridional circulation was without justification described as a purely diffusion process.

In 1997, the Zahn scheme was applied by Meynet and Maeder [135] in studies of the evolution of rotating MS stars with masses of 9, 20, 40 and 60 $M_\odot$, and by Talon *et al.* [196] in calculations of the influence of meridional circulation and turbulent diffusion on the abundance profiles of He and CNO-cycle elements in a MS star with $M = 9 M_\odot$. However, in these two studies, the transition of the star to a state of stationary rotation, in which the function $\Omega(r)$ evolves from the assumed initial state of uniform rotation to some asymptotic distribution, is not discussed. The authors simply used stationary solutions of Zahn's equation describing the redistribution of angular momentum, and followed only slow time variations associated with the evolution of the internal structure of the star on the main sequence.

In this section, we study the transition of a massive MS star to a state of stationary rotation. Since the relaxation time required for this transition proves to be very short compared to the life span of the stars on the main sequence, we can neglect variations in the H abundance. In this case, we can use a model zero-age MS star as a first approximation throughout the transition to the rotation state in an asymptotic regime. To elucidate how our results depend on the star's mass and rate of rotation, we conducted calculations for two masses, 10 and 30 $M_\odot$, and considered two surface rotation rates for the models with $M = 10\,M_\odot$.

4.2 Basic Equations

We introduced corrections to the expression for the amplitude of the radial component of the rate of meridional circulation U obtained by Zahn ([219], Eqs. (3.37–3.39)), in order to take into account the contribution of radiation pressure, which could be appreciable for massive MS stars. Following a procedure analogous to that used by Zahn, we found

$$U(r) = \frac{L}{Mg}\left(\frac{P}{c_P \rho T}\right)\frac{1}{\nabla_{\rm ad} - \nabla}(E_\Omega + E_\mu), \qquad (4.1)$$

where

$$\begin{aligned}
E_\Omega &= \frac{8}{3}\frac{\Omega^2 r^3}{GM}\left(1 - \frac{\Omega^2}{2\pi G\rho} - \frac{\varepsilon}{\varepsilon_{\rm m}}\right) \\
&\quad - \frac{\rho_{\rm m}}{\rho}\left\{\frac{r}{3}\frac{d}{dr}\left[H_T\frac{d(\frac{\Theta}{\delta})}{dr} - (\chi_T + 1 - \delta)\frac{\Theta}{\delta}\right] - 2\frac{H_T}{r}\frac{\Theta}{\delta} + \frac{2}{3}\Theta\right\} \\
&\quad - \frac{\varepsilon}{\varepsilon_{\rm m}}\left[H_T\frac{d(\frac{\Theta}{\delta})}{dr} + (\varepsilon_T - \chi_T - 1 + \delta)\frac{\Theta}{\delta}\right] \\
&\quad - \frac{\Omega^2}{2\pi G\rho}\Theta, \qquad (4.2)
\end{aligned}$$

and

$$E_\mu = \frac{\rho_m}{\rho}\left\{\frac{r}{3}\frac{d}{dr}\left[H_T\frac{d(\frac{\Lambda}{\delta})}{dr} - (\chi_T+1)\frac{\Lambda}{\delta} - \chi_\mu\Lambda\right] - 2\frac{H_T}{r}\frac{\Lambda}{\delta}\right\}$$

$$+ \frac{\varepsilon}{\varepsilon_m}\left[H_T\frac{d(\frac{\Lambda}{\delta})}{dr} + (\varepsilon_T - \chi_T - 1)\frac{\Lambda}{\delta} + (\varepsilon_\mu - \chi_\mu)\Lambda\right]. \quad (4.3)$$

In (4.1), L is the luminosity at the surface of a sphere of radius r, M the mass enclosed by this sphere (the Lagrangian mass co-ordinate), $g = GM/r^2$ the local gravitational acceleration, $P = (\mathcal{R}/\mu)\rho T + \frac{1}{3}aT^4$ the total pressure, c_P the specific heat capacity at constant pressure, $\nabla_{ad} = (d\ln T/d\ln P)_{ad}$ the adiabatic temperature gradient, and ∇ the actual gradient, which is assumed to be $\nabla_{rad} = 3/(16\pi Gac)(\kappa P/T^4)(L/M)$ in the radiative zone, where κ is the opacity. The remaining notation has the usual meanings. In expressions (4.2) and (4.3), we used the same notation for the measurement of horizontal density fluctuations $\Theta = \tilde{\rho}/\rho = (r^2 d\Omega^2)/(3gdr)$ and the mean molecular weight $\Lambda = \tilde{\mu}/\mu$ as in Zahn's paper [219]. H_T is the temperature scale height, and $\varepsilon_m = L/M$ and $\rho_m = M/(\frac{4\pi}{3}r^3)$ give the mean rate of energy production and mean density respectively. ε is the local rate of release of nuclear energy, $\chi = (4acT^3)/(3\kappa\rho)$ the coefficient of radiative heat conduction, and the symbols ε_T, χ_T, ε_μ and χ_μ are used to denote their logarithmic derivatives in T and μ. The quantity $\delta = -(\partial\ln\rho/\partial\ln T)_{\mu,P} = (4-3\beta)/\beta$, where $\beta = (\frac{\mathcal{R}}{\mu}\rho T)/P$, approaches unity when the ideal-gas pressure begins to dominate over the radiation pressure. In this limit, our equations (4.2) and (4.3) reduce to Zahn's equations (3.37) and (3.38) with the necessary small corrections to (3.37) (Zahn, private communication; see also [132]).

The evolution of the angular velocity profile is investigated by solving the following equation for the redistribution of angular momentum:

$$\frac{\partial}{\partial t}(\rho r^2 \Omega) = \frac{1}{5r^2}\frac{\partial}{\partial r}(\rho r^4 \Omega U) + \frac{1}{r^2}\frac{\partial}{\partial r}\left(\rho r^4 \nu_v \frac{\partial \Omega}{\partial r}\right), \qquad (4.4)$$

where ν_v is the vertical component of the turbulent viscosity [219, 220]. This is a non-linear equation in 4th-order partial derivatives. For ν_v, which is taken to coincide with the value of the coefficient of vertical turbulent diffusion D_v, we used the expression published in [196]:

$$\nu_v \approx D_v \approx \frac{8 Ri_c}{5} \frac{\left(r\dfrac{d\Omega}{dr}\right)^2}{N_T^2/(K + D_h) + N_\mu^2/D_h}. \qquad (4.5)$$

This was first derived by Talon and Zahn [195]. In (4.5), $Ri_c \approx 1/4$ is the critical Richardson number, $K = \chi/(c_P \rho)$ the coefficient of radiative-energy diffusion, $N_T^2 = (g/H_P)\delta(\nabla_{\rm ad} - \nabla)$ and $N_\mu^2 = (g/H_P)\varphi\nabla_\mu$ the squares of the T and μ components of the buoyancy frequency, H_P the pressure scale height and $\varphi = (\partial \ln \rho/\partial \ln \mu)_{P,T} = 1$ for a mixture of ideal gas and radiation. Unfortunately, it is not possible to derive an expression for the horizontal component of the coefficient of turbulent viscosity D_h from first principles. Zahn [219] suggested the following estimate:

$$D_h \approx \frac{rU}{C_h}\left[\frac{1}{3}\frac{d\ln(\rho r^2 U)}{d\ln r} - \frac{1}{2}\frac{d\ln(r^2\Omega)}{d\ln r}\right] \qquad (4.6)$$

with the free parameter $C_h \approx 1$. The structure of (4.6) explicitly takes into account the fact that precisely meridional circulation is responsible for the production and maintenance of a state of differential rotation in level surfaces of the star, so that D_h should probably be proportional to U.

The right-hand side of (4.4) contains two terms describing the transport of angular momentum by meridional-circulation flows and vertical turbulent diffusion. As a consequence of the horizontal erosion referred to above, the contribution of meridional circulation to the right-hand side of the nuclear kinetics equation is described by a diffusion term (for a more detailed explanation, see [40])

$$\frac{\partial y_i}{\partial t} = \left(\frac{\partial y_i}{\partial t}\right)_{\text{nucl}} + \frac{1}{\rho r^2}\frac{\partial}{\partial r}\left[\rho r^2 (D_{\text{v}} + D_{\text{eff}})\frac{\partial y_i}{\partial r}\right], \qquad (4.7)$$

where $(\partial y_i/\partial t)_{\text{nucl}}$ is the sum of the source and sink terms taking into account the contribution of nuclear reactions, y_i is the relative abundance of nuclide i (in terms of number of particles), and

$$D_{\text{eff}} = \frac{|rU|^2}{30D_{\text{h}}}. \qquad (4.8)$$

Equations (4.1)–(4.8) supplemented by the appropriate initial and boundary conditions (see the next section) form a closed system.

4.3 Additional Assumptions and Simplifications

As indicated in the Introduction, we used only model zero-age MS stars without considering their subsequent evolution, since the task in hand was to calculate the transition of the stars to a state of stationary rotation, which occurs rapidly compared to their MS life spans. We considered two models with masses $M = 10$ and $30\,M_\odot$ and abundances of hydrogen and heavy elements $X = 0.70$ and $Z = 0.02$. The models were calculated using an updated version of the code used earlier in [64]. The revised code is described in [167], and makes use of the most recent tables of opacities [157, 95] and thermodynamic functions [158]. All the required $\Omega(r)$-independent quantities in Eqs. (4.1)–(4.8) were calculated using the prepared models, after which they were no longer recalculated.

We did not introduce corrections in the equation of hydrostatic equilibrium to take into account the influence of rotation on the internal structure of the star, since Meynet and Maeder [135] indicate such corrections to be insignificant.

Assuming that convection can penetrate into the radiative envelope by a distance of order $0.1\,H_P$ relative to the formal boundary of the convective core r_{c}, defined by the condition $\nabla_{\text{rad}} = \nabla_{\text{ad}}$, we applied inner boundary conditions for Eq. (4.4) at $r = r_{\text{i}} \equiv r_{\text{c}} + 0.1 H_P$. This enabled us to remove a singularity when calculating U using

(4.1). The inner boundary conditions were taken to be

$$\begin{cases} \partial\Omega/\partial r = 0, & \text{and} \\ \Omega = \Omega_c, \end{cases} \qquad (4.9)$$

where

$$\frac{d\Omega_c}{dt} = \frac{(\rho r^4 \Omega U)_{r=r_c}}{5 \int_0^{r_c} \rho r^4 \, dr}. \qquad (4.10)$$

Relation (4.10) comes from the assumptions that *(i)* the convective core rotates rigidly as a consequence of the large values of the coefficient of (convective) turbulent viscosity in it, and *(ii)* the total angular momentum of the star remains constant; i.e. we do not allow for mass loss by the star (this assumption is discussed in more detail in the Conclusion).

The outer boundary conditions applied near the stellar surface were

$$\begin{cases} \partial\Omega/\partial r = 0, & \text{and} \\ U = 0. \end{cases} \qquad (4.11)$$

These follow directly from assumption *(ii)*.

As an initial condition, we assumed uniform rotation, $\Omega(r) = \Omega_0$, where a value of Ω_0 consistent with our assumptions was determined in accordance with the following scheme.

We first searched for an asymptotic solution to Eq. (4.4) for a specified angular velocity of rotation at the surface Ω_s by solving the stationary equation

$$-\frac{1}{5}\Omega U = \nu_v \frac{d\Omega}{dr}, \qquad (4.12)$$

which is obtained from (4.4) by setting the partial derivative with respect to time equal to zero [219, 201, 196]. The boundary value problem for the ordinary differential equation (4.12) (the boundary conditions reduce to the first lines of (4.9) and (4.11), supplemented by the requirement that $\Omega = \Omega_s$ at the surface) was solved using a "shooting" method. Further, Ω_0 was chosen so as to ensure conservation of the total angular momentum of the star.

DEEP MIXING IN GLOBULAR-CLUSTER RED GIANTS   151

To simplify the calculations, we assumed that the radiative energy diffusion coefficient K was much larger than the coefficient of horizontal turbulent diffusion D_h. Another simplification was that we neglected the release of nuclear energy and the gradient of μ in the radiative envelope. Recall that we did not consider the evolution of the star, so that our "radiative envelope" was actually the region from the boundary of the convective core at its maximum size (on the zero-age main sequence) to the stellar surface. Standard stellar-evolution calculations show that during the life span of a massive star on the main sequence, its convective core decreases in size, leaving the zone with variable chemical composition in which semi-convection can arise outside its boundary. We do not discuss the rather non-trivial problem of how extra-mixing can overcome the barrier created by the μ gradient, if, indeed, it can (see [132]). Instead, we strive to answer the question of whether extra-mixing will be sufficiently rapid for products of nuclear reactions to be carried to the stellar surface from layers adjacent to the boundary of the convective core at a radius r_i during the star's life on the main sequence. Our second assumption is admissible in the framework of this problem.

Our two simplifying assumptions enable us to set $E_\mu = 0$ and $N_\mu^2 = 0$, and to use the approximate formula

$$\nu_v \approx D_v \approx \frac{8Ri_c}{5} \frac{K}{N_T^2} \left(r \frac{d\Omega}{dr} \right)^2 \qquad (4.13)$$

in place of (4.5).

The solutions of the total non-stationary problem (Eqs. (4.1)–(4.4), (4.6), (4.8)–(4.11) and (4.13)) were obtained using the Henyey method.

Before turning to a discussion of the results of our calculations, we will note the criteria for self-consistency that we adopted as a condition for applicability of our Zahn scheme. These criteria will be verified *a posteriori*.

$$D_v \geq \frac{1}{3} \nu Re_c, \qquad (4.14)$$

where ν is the usual (i.e. molecular plus radiative) viscosity and Re_c is the critical Reynolds number. If (4.14) is not satisfied, the motion that is responsible for transport via vertical turbulent diffusion will be suppressed by viscous friction.

The second strong inequality

$$D_\mathrm{v} \ll K \tag{4.15}$$

must be satisfied if we wish to neglect the transport of heat by turbulence.

$$K \gg D_\mathrm{h} \tag{4.16}$$

is one of our simplifying assumptions, which, together with the assumption of zero gradient in μ, transforms (4.5) into (4.13).

$$D_\mathrm{v} \ll D_\mathrm{h}. \tag{4.17}$$

This is Zahn's main assumption, which enables a one-dimensional treatment of the problem.

4.4 Calculation Results

Figure 4.1 presents solutions of the non-stationary problem for the stellar model with $M = 30\,M_\odot$. We adopted $V_\mathrm{s} = 470$ km·s^{-1} for the rate of rotation at the surface, which is close to the maximum value observed for OB stars. This corresponds to an angular velocity $\Omega_\mathrm{s} = 10^{-4}$ rad·s^{-1} and a ratio of the centrifugal acceleration to the gravitational acceleration at the equator $\eta_\mathrm{s} = \Omega_\mathrm{s}^2 R^3/GM \approx 0.26$. The solid curve in Fig. 4.1a gives the stationary asymptotic solution (Eq. (4.12)). We can see that in this specific case, the non-stationary solution approaches the stationary solution very rapidly (the numbers near the curves correspond to ages in years). The life span of the star with $M = 30\,M_\odot$ on the main sequence is $\tau_\mathrm{MS} = 5.9 \cdot 10^6$ yrs. Thus, when $\Omega_\mathrm{s} = 10^{-4}$ rad·s^{-1} the star requires only about 1% of its life span on the main sequence to make the transition to a state of stationary rotation. This relaxation time can be approximated *a priori* as the ratio $\tau_\mathrm{rel} \approx R/U(\Omega_\mathrm{s})$ (see below), which, for a fixed mass, is roughly proportional to Ω_s^{-2} (Eqs. (4.1)–(4.2)). It is obvious that the more rapidly the star rotates, the more quickly it reaches the stationary state.

DEEP MIXING IN GLOBULAR-CLUSTER RED GIANTS

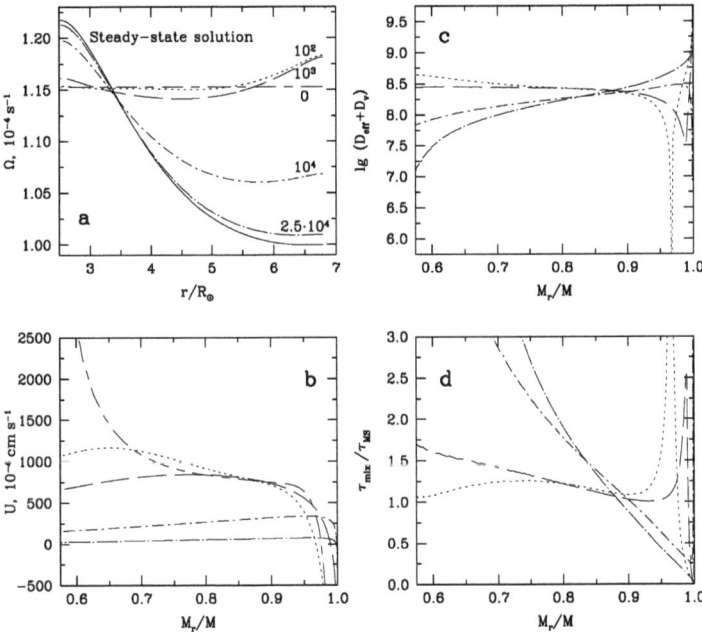

Figure 4.1. Approach to a state of stationary rotation for a star with $M = 30\,M_\odot$ with surface rotational velocity $V_s = 470$ km·s^{-1} (corresponding to $\Omega_s = 10^{-4}$ rad·s^{-1}). The stationary asymptotic solution is depicted by the solid curve in panel **a**. The numbers near the curves in **a** give the ages (in years) of the corresponding angular velocity profiles. Panels **b**, **c** and **d** show the evolution of the rate of meridional circulation, total diffusion coefficient and ratio of the characteristic mixing time $\tau_{\rm mix} = (R-r)^2/(D_{\rm eff} + D_{\rm v})$ to the MS life span, respectively. In all panels, curves of the same type have the same age. In this and subsequent figures in which the co-ordinate $r/R_\odot$ (or M_r/M) is plotted on the horizontal axis, these quantities are always measured from the left, approximately from the boundary of the convective core or, more precisely, from the point where $r = r_{\rm c} + 0.1 H_P$ (see text for further explanation).

Panels b and c of Fig. 4.1 show the evolution of the distributions of the rate of meridional circulation and the total diffusion coefficient. When the initial rotation is uniform, U changes sign near the surface. This well-known result is due to the negative term $-\Omega^2/2\pi G\rho$ in parantheses in the first line of expression (4.2) [85, 140]. In a classical description with uniform rotation, the meridional-circulation flows are enclosed in two zones that are isolated from each other, with the mass of the outer zone being much lower than that of the inner zone [144]. Slow mixing between these zones occurs only via microscopic diffusion [44]. In the past, models proposing that the boundary between these zones should prevent total mixing in the radiative envelope of a single rotating MS star were considered [204, 122]. However, as our calculations show (see also [201, 196]), in the Zahn scheme, the boundary between these zones shifts toward the surface as the rotation profile evolves. As a result, the outer zone gradually (but for sufficiently high values of Ω_s, quite rapidly) contains less and less mass, finally disappearing completely. Simultaneously, the characteristic value of U (say, in the middle of the envelope) decreases by a factor of about 50. While the rotation profile remains close to the initial flat profile, D_{eff} (i.e. the meridional circulation) makes the defining contribution to the total diffusion coefficient. When $\Omega(r)$ approaches its asymptotic distribution, D_v (i.e. turbulent diffusion) begins to dominate. Overall, in the transition to the stationary regime, the sum $D_{\text{eff}} + D_v$ decreases near the boundary of the convective core by a factor of about $10^{1.5}$ compared to the case of uniform rotation (Fig. 4.1c).

In Fig. 4.1d, we plot the ratio $\tau_{\text{mix}}/\tau_{\text{MS}}$ as a function of the relative mass co-ordinate. The mixing time here is defined as $\tau_{\text{mix}} = (R - r)^2/(D_{\text{eff}} + D_v)$. We can see that even for rotation profiles that are nearly flat, τ_{mix} exceeds τ_{MS} in the bulk of the radiative envelope. This contrasts strongly with the classical (Eddington–Sweet) estimate of the mixing time

$$\tau_{\text{ES}} \approx \frac{\tau_{\text{KH}}}{\eta_s},$$

where $\tau_{\text{KH}} = GM^2/RL$ is the Kelvin–Helmholtz timescale. For our model zero-age MS star with $M = 30\,M_\odot$, we have $\tau_{\text{KH}} \approx 3.4 \cdot 10^4$ yrs, so that with $\eta_s = 0.26$ and $\tau_{\text{MS}} = 5.9 \cdot 10^6$ yrs, the classical estimate gives $\tau_{\text{mix}}/\tau_{\text{MS}} \approx 0.02$! The main reason for this large discrepancy in the mixing times (for the Zahn scheme and a classical description) is horizontal erosion, which is responsible for the factor of $|rU|/30D_h \ll 1$ in expression (4.8), which appreciably decreases the product $|rU|$ corresponding to the classical estimate of the rate of meridional-circulation mixing. Precisely this factor leads to a low value for the coefficient D_{eff}, even for rotation that is nearly uniform.

Note that in evolutionary stellar models, circulatory flows and diffusion never reach a state in which they precisely compensate each other, making the time derivative in (4.4) equal to zero. Therefore, it is correct to think of τ_{rel} not as the time to reach the stationary regime (which in reality never happens), but as the characteristic time for variations of the inner rotation profile.

Figure 4.2 presents the same distributions of parameters as in Fig. 4.1 for model stars with $M = 10\,M_\odot$ rotating with a surface velocity $V_s = 230$ km·s^{-1} ($\Omega_s = 9 \cdot 10^{-5}$ rad·s^{-1}). For this mass, we also carried out calculations with $V_s = 460$ km·s^{-1} ($\Omega_s = 1.8 \cdot 10^{-4}$ rad·s^{-1}).

A comparison of the calculation results for the two selected stellar masses and two rotation rates for the model with $M = 10\,M_\odot$ led to the following crude estimate of the relaxation time:

$$\tau_{\text{rel}} \approx R/U(\Omega_s) \approx \tau_{\text{ES}} \propto \tau_{\text{KH}} \Omega_s^{-2} \left(\frac{M^2}{R^3} \right). \qquad (4.18)$$

This is much shorter than the mixing time τ_{mix} above, since the transport of angular momentum by meridional circulation is not subject to horizontal erosion (Eq. (4.4)). Precisely this difference in the mixing rates of chemical elements and the transport rates of angular momentum is considered to be one of the main advantages of the Zahn scheme, since this can explain the fairly rapid slowing of the rotation of the Sun and similar stars, accompanied by a very gradual decrease in their surface abundances of Li [220].

If we use the mass–radius relation from [7], $R \propto M^{0.52}$, valid for MS stars with masses $15 \leq M/M_\odot \leq 60$, then $M^2/R^3 \propto M^{0.44}$, so that to first approximation, $\tau_{\text{rel}} \propto \tau_{\text{KH}} \Omega_s^2$. This relation is satisfied (within a factor of 3–4) in our numerical calculations. For example, the model with $M = 10\,M_\odot$ has $\tau_{\text{KH}} \approx 1.5 \cdot 10^5$ yrs (the Kelvin–Helmholtz timescale increases with decreasing M on the main

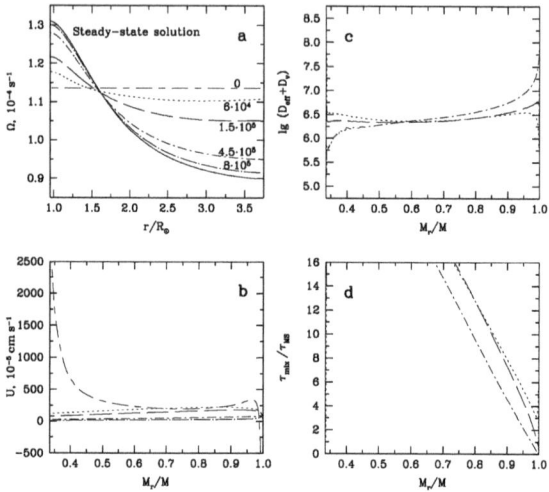

Figure 4.2. Same as Fig. 4.1 for a star with $M = 10\,M_\odot$ and $V_s = 230$ km·s^{-1} ($\Omega_s = 9 \cdot 10^{-5}$ rad·s^{-1}).

sequence), and, with $\Omega_s = 1.8 \cdot 10^{-4}$ and $9 \cdot 10^{-5}$ rad·s^{-1} exact calculations give $\tau_{\rm rel} = 3.4 \cdot 10^4$ and 10^6 yrs, respectively (compare with the values $\tau_{\rm KH} \approx 3.4 \cdot 10^4$ and $\tau_{\rm rel} = 5 \cdot 10^4$ yrs for the model star with $M = 30\,M_\odot$ rotating with velocity $\Omega_s = 10^{-4}$ rad·s^{-1}).

We found that the rate of meridional circulation and the total diffusion coefficient vary approximately in proportion to Ω_s^2; for example, when $\Omega_s = 1.8 \cdot 10^{-4}$ rad·s^{-1} the dependence of $\lg(D_{\rm eff} + D_{\rm v})$ on M_r/M lies roughly 0.6 ($\approx \lg 4$) above the curves depicted in Fig. 4.2c for the case $\Omega_s = 9 \cdot 10^{-5}$ rad·s^{-1}.

We can conclude from Figs. 4.2c and 4.2d that, not considering very deep layers adjacent to the convective core, there are no major differences between rotation states that are close to uniform and that approach the asymptotic state in terms of their ability (or more accurately their inability) to mix elements. In particular, total mixing of the radiative envelope does not come about in a single case. Therefore, the search for other angular momentum transport mechanisms (such as internal gravitational waves [221]) that could maintain a state of nearly uniform rotation are unlikely to help accelerate the mixing of elements – if, of course, the new mechanism itself cannot efficiently perform this mixing. However, even in the most favourable of the cases we have considered, with $M = 10\,M_\odot$, $\Omega_s = 9 \cdot 10^{-5}$ rad·s^{-1} and $t = 8 \cdot 10^5$ yrs, turbulent diffusion (recall that D_eff is much lower than D_v for rotation that is close to stationary) is sufficiently rapid for the production of *some* mixing, leading to appreciable variations in the surface abundance of ^{14}N by the end of the star's life span on the main sequence (Fig. 4.3). The curves in Fig. 4.3 were obtained by solving Eq. (4.7) using the method and input physical data described in [66] (using a system of kinetics equations for 26 particles). The required T and ρ distributions were taken from our models for zero-age MS stars. We emphasise that the surface abundance of N began to grow only after some time delay, corresponding to the time required for the "wave" of diffusion to reach the stellar surface (Fig. 4.3b, see also [196]).

Figure 4.4 demonstrates that soon after the rotation profile begins to evolve, all the self-consistency criteria (4.14)–(4.17) begin to be satisifed; only in the very beginning of the evolution did we have $K \leq D_h$, but we have neglected this.

4.5 Main Conclusions

We have considered the evolution of stellar rotation profiles associated with the redistribution of angular momentum by meridional circulation and turbulent diffusion in the radiative zones of model MS stars with masses of 10 and 30 $M_\odot$. We have applied the main assumptions and equations of Zahn [219] and Talon and Zahn [195].

Note that we have not taken into account the gradient in the mean molecular weight in our calculations. This is justified by the fact that the resulting timescale for the transition of the stars to a stationary rotation state is much shorter than their life span on the main sequence.

One of our simplifying assumptions was to neglect mass loss by the star. We were essentialy forced to make this assumption, since

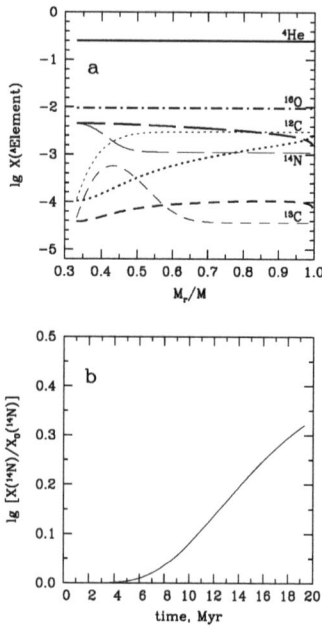

Figure 4.3. Panel **a**: Distributions of abundances ($X_i = y_i A_i$, A_i is the atomic mass number) of main CNO-cycle nuclides and He in the radiative envelope of a star with $M = 10 M_\odot$ toward the end of its life on the main sequence, calculated for two cases: without mixing (thin curves) and with mixing by turbulent diffusion due to rotation with a surface velocity $V_s = 230$ km·s^{-1} (thick curves; here, meridional circulation does not play a major role). The profile of the diffusion coefficient in the calculations with mixing was taken from the non-stationary solution for age $t = 8 \cdot 10^5$ yrs (Fig. 4.2a). Panel **b**: Increase in the surface abundance of N with time in the mixing calculations. The N abundance begins to increase only after some time delay, corresponding to the time required by the wave of diffusion to reach the surface.

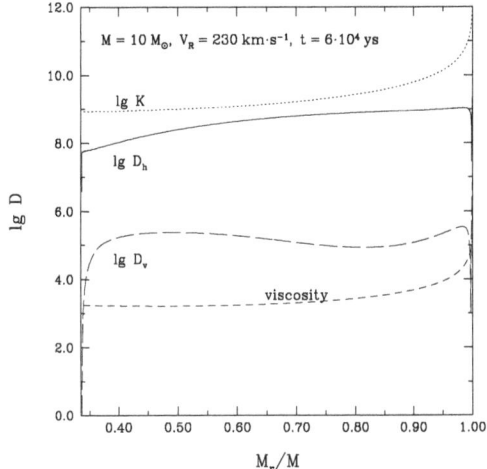

Figure 4.4. Illustration that the self-consistency criteria (4.14)–(4.17) begin to be satisifed immediately after the beginning of the evolution of the rotation profile.

it remains unclear what boundary conditions for Eq. (4.4) should be applied in the presence of a stellar wind. Semi-empirical formulae for the mass-loss rates of MS stars can be used only to calculate the angular momentum loss rate; i.e. they essentially provide outer boundary conditions for the integral of Eq. (4.4). However, even here it is not clear what fractions of the angular momentum are carried to the atmosphere of the star by meridional circulation and by turbulent diffusion. Here, we can only hope that we are dealing with the simple case with "the dog wagging its tail" and not the other way around, so that our neglect of the stellar wind does not mean that we have "thrown the baby out with the bathwater".

Based on the results of our calculations, we arrived at the following conclusions.

1. With sufficiently high values of the angular rotational velocity at the surface Ω_s, which can be estimated *a priori*, the relaxation time τ_{rel} required for the transition of a massive star to a state of stationary rotation proves to be much less than its life span on the main sequence τ_{MS}. Since our estimate yields $\tau_{rel} \ll \tau_{MS}$, the assumption that the star is in a state of stationary rotation from the very beginning of its evolution on the main sequence, used by Talon *et al.* [196], is admissible. On the other hand, at sufficiently low values of Ω_s, this assumption is not correct, and the evolution of the rotation profile must be traced by solving Eq. (4.4) in parallel with stellar-evolution calculations.

2. Qualitatively, the non-stationary solutions do not greatly differ from the stationary solutions in terms of the ability of the corresponding meridional circulation and turbulent diffusion to mix chemical elements. Even with nearly uniform rotation, the rate of extra-mixing (in this case, provided by meridional circulation) is much lower than the classical estimate $|rU|$. The reason for this is the erosion of inhomogeneities in the chemical composition in surface layers produced by horizontal turbulent diffusion. Complete mixing in the radiative envelope (the establishment of uniform distributions of elemental abundances) did not occur in any of the cases we considered.

3. In spite of the low mixing rate, in the presence of sufficiently high values of Ω_s, turbulent diffusion (which dominates over meridional circulation when the rotation is close to stationary) is able to appreciably change the surface abundances of N and C in the stars by the time they leave the main sequence. An important result here is that this occurs with some time delay (up to 50% of the life span of the star on the main sequence), during which the "wave" of diffusion reaches the surface and the abundance variations become accessible to observations (Fig. 4.3). It is interesting that Lyubimkov [126] reported a similar delay in the appearance of He excesses in the atmospheres of OB

stars. If extra-mixing is able to penetrate into the convective core which is most likely in the early stages of evolution of the star on the main sequence [58]), the evolution of the surface abundances of He in massive MS stars could be similar to the time dependence of the N abundance depicted in Fig. 4.3.

5. CONCLUSION

The studies discussed in this review have provided further support and development for the theoretical interpretation of the relationship between the abundances of Na and CNO-cycle elements in globular-cluster red giants proposed in 1990 [61]. The essence of this interpretation is as follows.

In the outer layers of the hydrogen burning shell in a red giant, the temperature is sufficiently high for the synthesis of Na (in the reaction $^{22}Ne(p,\gamma)^{23}Na$) and Al to occur, together with decreases in the C and O abundances and an increase in the N abundance associated with the CNO cycle – the main source of energy in red giants. Depending on the MgAl-cycle reaction rates (which remain uncertain), aluminium could be produced either as the stable isotope ^{27}Al, in the reaction chain $^{25}Mg(p,\gamma)^{26}Al^g(p,\gamma)^{27}Si(\beta^+\nu)^{27}Al$ (if the rate of the reaction $^{26}Al^g(p,\gamma)^{27}Si$ is sufficiently high), or as the radioactive isotope $^{26}Al^g$, in the reaction $^{25}Mg(p,\gamma)^{26}Al^g$. The idea that N and Na (and now Al as well) produced in globular-cluster red giants is transported outward to the base of the convective envelope by some form of extra-mixing was first proposed in [61]). The scale of the variations in the surface chemical composition of red giants depends on the effiicency of the extra-mixing, which can vary from star to star (for reasons that remain unknown).

We have proposed two models for extra-mixing in red giants: *(i)* a semi-empirical diffusion model with the depth and rate of the mixing chosen to fit the observations (see [64] and Section 2.1); and *(ii)* the first (and as yet the only) real physical model ([63] and Section 2.2), based on the Zahn mechanism with meridional circulation and turbulent diffusion [219, 132]. Both models can reproduce the observed (anti)correlations between abundances of CNO-cycle elements and Na and Al in red giants in ω Cen and M4 fairly well, as well as the global anticorrelation of [O/Fe] with [Na/Fe] (Figs. 3.2, 2.12, 2.13, 3.10, 3.11).

In the Zahn mechanism, extra-mixing is brought about by the star's rotation. Indirect confirmation that rotation has some relationship to the origin of chemical composition anomalies in red giants is provided by the observational facts that: *(i)* horizontal-branch stars with unusually rapid rotation are found in the globular cluster M13, which also contains giants with extremely low O abundances [145]; *(ii)* on average, field giants with lithium excesses rotate more rapidly than normal K giants (N. A. Drake, private communication).

In extra-mixing scenarios, the anomalously high Al overabundances in red giants in ω Cen, M4 and M13 require that the initial abundance of ^{25}Mg exceeded a value proportional to the solar abundance by more than an order of magnitude.

In [66] (see also Section 3.1), we first showed that the required excess of ^{25}Mg could be produced in intermediate-mass AGB stars formed in generations preceding the currently observed red giants; i.e. in "primordial nucleosynthesis" processes in globular clusters.

Since the publication of [61], a whole series of new observational data has provided evidence for the presence of variations in the abundances of C, N, Na, and possibly even O, Al and Mg, in stars near the main sequences of some globular clusters. This brought about a need to combine scenarios with extra-mixing and with primordial nucleosynthesis. The first such combined scenario was proposed in [66]; subsequent modification of this scenario is discussed in Section 3.2.

Based on this combined scenario, we propose the following *model for the chemical evolution of globular clusters.*

The evolution of the chemical composition of globular clusters apparently differs significantly from the chemical evolution of the Galaxy as a whole [199]. One probable scenario for the formation of globular cluster was suggested by Cayrel [39], and later developed quantitatively by Brown et al. [26] and Parmentier et al. [143]. These two studies supplemented Cayrel's scenario in two important ways: they showed first, that a protocluster could experience hundreds of type-II supernovae and not be disrupted and second, that the entire range of [Fe/H] values observed in globular clusters could be obtained in the framework of the scenario. Cayrel's scenario forms the basis for our model described below.

DEEP MIXING IN GLOBULAR-CLUSTER RED GIANTS   163

We suggest that after a massive protocluster cloud has become distinct from the protogalactic medium, at first only massive stars (we will refer to them as the "first generation") are formed in the dense core of material with the cosmological chemical composition ($Z = 0$). These stars evolve rather rapidly and explode as type-II supernovae; for some reason (see, for example, the work of Nakamura and Umemura [136]), low-mass stars are not formed at this stage (indeed, they are not observed!). The supershell created by the shock waves from the numerous supernova explosions sweeps up and compresses the protocluster material, which now begins to consist of a mixture of matter with the cosmological composition and matter ejected by the supernovae.

We believe that this material then forms stars with the entire spectrum of admissible masses (we refer to these as the "second generation" of globular-cluster stars). The most massive of these stars also become type-II supernovae, whose shock waves clear the cluster of a large fraction (if not all) of the remaining gas. The matter lost by second-generation, intermediate-mass AGB stars has appreciably lower kinetic energy than that ejected by supernovae, so that it remains in the interstellar medium (possibly mixing with the remnants of the older material).

Finally, a "third generation" of stars may form from this new mixture of material. Note that the matter lost by second-generation AGB stars has the same [Fe/H] as the matter from which they formed: it is known (and we have shown) that only the abundances of CNO nuclides, the Mg isotopic ratios, the ^{22}Ne/Na ratio, and a number of other abundances vary in AGB stars. Thus, the currently observed stars in globular clusters could belong to either the second or the third generation. Their observed or inferred chemical composition anomalies (such as an overabundance of ^{25}Mg) were either present in the material from which they formed or were acquired by the stars in the course of their evolution, as a result of accretion of material from the interstellar medium and the action of deep mixing on the red giant branch.

Additional observational evidence supporting our proposed model for the chemical evolution of globular clusters includes the results of the following studies. *(i)* In [98], signs of nucleosynthesis occurring in type-II supernovae, and then in AGB stars as well (with their relative contribution correlated with the metallicity, [Fe/H]), were detected in metal-poor field dwarfs. It was proposed that these dwarfs were former members of globular clusters. *(ii)* In [96], an overabundance of barium, [Ba/Fe] = +0.60, was discovered in red giants in M4. This suggests an enrichment of the (proto)cluster medium in elements formed in the s-process, which is known to occur precisely in AGB stars.

The time between the formation of stars of the second and third generations corresponds to the characteristic life span of intermediate mass AGB stars, i.e. $\sim 10^8$ yrs, which is negligible compared to the ages of modern globular-cluster red giants. Consequently, all the stars in an individual globular cluster should have the same [Fe/H] values, but at least some of them could display excesses of ^{25}Mg and the other abundance peculiarities discussed above (an exception is ω Cen, in which the products of second-generation supernova explosions could be retained due to the cluster's unusually high mass, leading to the observed variations in [Fe/H]). The details of this picture will depend on the star formation rate, initial mass function, etc., but the proposed model is qualitatively able to explain *all* the observational data on the chemical compositions of globular-cluster stars in a self-consistent way.

As shown in [67] (see Section 4.1), the Zahn mixing mechanism could also be responsible for the characteristic chemical compositions of the atmospheres of OB stars. Namely, due to the quasi-stationary conditions in the radiative envelopes of massive MS stars, the actions of meridional circulation and turbulent diffusion almost completely compensate each other after a short (thermal) time. As a consequence, and also due to the horizontal erosion of the chemical composition by turbulence, extra-mixing in OB stars is very slow, in spite of their rapid rotation. This may explain the delay (up to 50% of the life span of OB stars on the main sequence) in the appearance of products of the CNO cycle in these stars' atmospheres indicated by some observations [126].

Our hypothesis concerning deep mixing in globular-cluster red giants [61] attracted the attention of spectroscopists in the USA (Lick and MacDonald Observatories), Australia (Mount Stromlo and Siding Spring Observatories), and, recently, Europe (European Southern Observatory), who initiated extensive observational programmes

DEEP MIXING IN GLOBULAR-CLUSTER RED GIANTS

aimed at elucidating the origin of the N, Na and Al excesses and C, O (and Mg) deficits in globular-cluster red giants – extra-mixing in these stars or inherited anomalies in chemical composition that arose in primordial nucleosynthesis. These studies involved the largest ground-based instruments available, such as the Keck and Kueyen (VLT UT2) telescopes.

Currently, it is fair to say that this goal has largely been achieved, and the solution of this problem is a scenario *combining* extra-mixing in red giants and primordial nucleosynthesis (as reported at the 24th IAU General Assembly; see materials for Joint Discussion 5, dedicated to mixing in stars).

This combined scenario puts forward new and topical tasks. On the theoretical side, it requires studies of the influence of extra-mixing on the red giant branch (for example, in the framework of the Zahn mechanism), on the position and evolution of horizontal branch stars, searches for other mechanisms for extra-mixing in red giants (for example, associated with the interaction of magnetic fields and rotation), quantitative analyses of the proposed model for the chemical evolution of globular clusters, etc. The primary observational tasks are analyses of the chemical compositions of stars near the main sequences of globular clusters and determinations of the isotopic composition of Mg for globular-cluster red giants (distinguishing between the isotopes ^{25}Mg and ^{26}Mg).

For the latest observational and theoretical results related to the problem of abundance inhomogeneities in globular clusters, the reader is referred to [223–234].

REFERENCES

1. Alongi, M., Bertelli, G., Bressan, A. and Chiosi, C., 1991, *A&A* **244**, 95.
2. Anders, E. and Grevesse, N., 1989, *Geochim. Cosmochim. Acta* **53**, 197.
3. Angulo, C. *et al.*, 1999, *Nucl. Phys. A* **656**, 3.
4. Arnould, M., Goriely, S. and Jorissen, A., 1999, *A&A* **347**, 572.
5. Arnould, M., Mowlavi, N. and Champagne, A., 1995, in: Noels, A. *et al.* (eds.), *Stellar Evolution: What Should Be Done*, Universite de Liège, Belgium, 1996, pp. 17–29.
6. Bao, Z. and Käppeler, F., 1987, *AD&NDT* **36**, 411.
7. Beech, M. and Mitalas, R., 1994, *ApJS* **95**, 517.

8. Beer, H., Käppeler, F. and Arcoragi, J.-P., in: Hillebrandt, W. and Müller, E. (eds.), *Proc. of the 5th Workshop on Nuclear Astrophysics*, p. 10, 1989.
9. Bell, R.A., Dickens, J.A. and Gustafsson, B., 1979, *ApJ* **229**, 604.
10. Bellman, S., Briley, M.M., Smith, G.H. and Claver, C.F., 2001, *PASP* **113**, 326.
11. Berdyugina, S.K. and Savanov, I.S., 1994, *SvA Lett.* **20**, 639.
12. Bernasconi, P.A. and Maeder, A., 1996, *A&A* **307**, 829.
13. Bonifacio, P. and Molaro, P., 1997, *MNRAS* **285**, 847.
14. Boothroyd, A.I. and Sackmann, I.-J., 1999, *ApJ* **510**, 232.
15. Boothroyd, A.I., Sackmann, I.-J. and Wasserburg, G.J., 1994, *ApJ* **430**, L77.
16. Boothroyd, A.I., Sackmann, I.-J. and Wasserburg, G.J., 1995, *ApJ* **442**, L21.
17. Briley, M.M., Bell, R.A., Hoban, S. and Dickens, R.J., 1990, *ApJ* **359**, 307.
18. Briley, M.M., Hesser, J.E., Bell, R.A., Bolte, M. and Smith, G.H., 1994, *AJ* **108**, 2183.
19. Briley, M.M., Smith, V.V., and Lambert, D.L., 1994, *ApJ* **424**, L119.
20. Briley, M.M., Smith, V.V., Suntzeff, N.B., Lambert, D.L., Bell, R.A. and Hesser, J.E., 1996, *Nature* **383**, 604.
21. Brown, J.A. and Wallerstein, G., 1989, *AJ* **98**, 1643.
22. Brown, J.A. and Wallerstein, G., 1989, *AJ* **106**, 133.
23. Brown, J.A., Sneden, C., Lambert, D.L. and Dutchover, E., 1989, *ApJS* **71**, 293.
24. Brown, J.A., Wallerstein, G., and Oke, J.B., 1991, *AJ* **101**, 1693.
25. Brown, J.H., Burkert, A. and Truran, J.W., 1991, *ApJ* **376**, 115.
26. Brown, J.H., Burkert, A. and Truran, J.W., 1995, *ApJ* **440**, 666.
27. Busso, M., Picchio, G., Gallino, R., Chieffi, A., 1988, *ApJ* **326**, 196.
28. Cameron, A.G.W., 1955, *ApJ* **121**, 144.
29. Cameron, A.G.W. and Fowler, W.A., 1971, *ApJ* **164**, 111.
30. Cameron, A.G.W., in: Barnes, C.A., Clayton, D.D. and Schramm, D.N. (eds.), *Essays in Nuclear Astrophysics*, Cambridge Univ. Press, Cambridge 1982, Chapter 3.
31. Cannon, R.D., Croke, B.F.W., Bell, R.A., Hesser, J.E. and Stathakis, R.A., 1998, *MNRAS* **298**, 601.
32. Carbon, D.F., Langer, G.E., Butler, D., Kraft, R.P., Suntzeff, N.B., Kemper, E., Trefzger, Ch.F. and Romanishin, W., 1982, *ApJS* **49**, 207.
33. Carney, B.W., Fry, A.M. and Gonzalez, G., 1998, *AJ* **116**, 2984.
34. Carney, B.W., in: Morrison, H. and Sarajedini, A. (eds.), *Formation of the Galactic Halo... Inside and Out*, ASP Conference Series, 1992, volume 92, p. 103.
35. Castilho, B.V., Spite, F., Barbuy, B., Spite, M., De Medeiros, J.R. and Gregorio-Hetem, J., 1999, *A&A* **345**, 249.
36. Caughlan, G.R. and Fowler, W.A., 1988, *AD&NDT* **40**, 283.
37. Cavallo, R.M. and Nagar, N.M., 2000, *AJ* **120**, 1364.
38. Cavallo, R.M., Sweigart, A.V. and Bell, R.A., 1996, *ApJ* **464**, L79.

39. Cayrel, R., 1986, A&A **168**, 81.
40. Chaboyer, B. and Zahn, J.-P., 1992, A&A **253**, 173.
41. Chandrasekhar, S., 1961, *Hydrodynamic and Hydromagnetic Stability*, Clarendon Press, Oxford, p. 491.
42. Charbonnel, C., 1994, A&A **282**, 811.
43. Charbonnel, C., 1995, ApJ **453**, L41.
44. Charbonnel, C. and Vauclair, S., 1992, A&A **265**, 55.
45. Charbonnel, C., Brown, J.A. and Wallerstein, G., 1998, A&A **332**, 204.
46. Cohen, J.G., 1978, ApJ **223**, 487.
47. Cohen, J.G., 1999, AJ **117**, 2428.
48. Cohen, J.G., 1999, AJ **117**, 2434.
49. Cottrell, P.L. and Da Costa, G.S., 1981, ApJ **245**, L79.
50. Cowan, J.J., Thielemann, F.-K. and Truran, J.W., 1991, *Phys. Rep.* **208**, 267.
51. Cox, J.P. and Guili, R.T., *Principles of Stellar Structure, Vol. I*, Gordon and Breach, New York, 1968, Chap. 14.
52. Da Costa, G.S., in: Bedding T.R. *et al.* (eds.), *Fundamental Stellar Properties: The Interaction Between Observations and Theory*, IAU Symp. 1997, vol. 189, p. 193.
53. Da Costa, G.S. and Cottrell, P.L., 1980, ApJ **236**, L83.
54. Da Silva, L., De la Reza, R. and Barbuy, B., 1995, ApJ **448**, L41.
55. De la Reza, R., Drake, N.A. and Da Silva, L., 1996, ApJ **456**, L115.
56. De la Reza, R., Drake, N.A., Da Silva, L., Torres, C.A.O. and Martin, E.L., 1997, ApJ **482**, L77.
57. Denissenkov, P.A., 1989, *Astrofizika* **31**, 293; 1990, *Astrophysics* **31**, 588.
58. Denissenkov, P.A., 1994, A&A **287**, 113.
59. Denissenkov, P.A., 1994, *Space Sci. Rev.* **66**, 405.
60. Denissenkov, P.A. and Denissenkova, S.N., 1989, *Astron. Tsirk.* **1538**, 11.
61. Denissenkov, P.A. and Denissenkova, S.N., 1990, *Pis'ma Astron. Zh.* **16**, 642; 1990, *Sov. Astronomy Letters* **16**, 275.
62. Denissenkov, P.A. and Ivanov, V.V., 1987, *Pis'ma Astron. Zh.* **13**, 214; 1987, *Sov. Astronomy Letters* **13**, 214.
63. Denissenkov, P.A. and Tout, C.A., 2000, MNRAS **316**, 395.
64. Denissenkov, P.A. and Weiss, A., 1996, A&A **308**, 773.
65. Denissenkov, P.A. and Weiss, A., 2000, A&A **358**, L49.
66. Denissenkov, P.A., Da Costa, G.S., Norris, J.E. and Weiss, A., 1998, A&A **333**, 926.
67. Denissenkov, P.A., Ivanova, N.S. and Weiss, A., 1999, A&A **341**, 181.
68. Denissenkov, P.A., Weiss, A. and Wagenhuber, J., 1997, A&A **320**, 115.
69. Dickens, R.J., Bell, R.A. and Gustafsson, B., 1979, ApJ **232**, 428.
70. Dickens, R.J., Croke, B.F.W., Cannon, R.D. and Bell, R.A., 1991, *Nature* **351**, 212.
71. Eddington, A.S., 1925, *Observatory* **48**, 78.
72. El Eid, M.F. and Champagne, A.E., 1995, ApJ **451**, 298.
73. Endal, A.S. and Sofia, S., 1976, ApJ **210**, 184.

74. Fall, S.M. and Rees, M.J., 1985, *ApJ* **298**, 18.
75. Fliegner, J., Langer, N. and Venn, K.A., 1996, *A&A* **308**, L13.
76. Fowler, W.A., Caughlan, G.R. and Zimmerman, B.A., 1967, *ARA&A* **5**, 525.
77. Freeman, K.C., in: Smith, G.H. and Brodie, J.P. (eds.), *The Globular Cluster-Galaxy Connection*, ASP Conference Series, 1993, vol. 48, p. 608, Santa Cruz, San Francisco.
78. Frost, C.A. and Lattanzio, J.C., 1996, *ApJ* **473**, 383.
79. Fujimoto, M.Y., Aikawa, M. and Kato, K., 1999, *ApJ* **519**, 733.
80. Gallino, R., Busso, M., Picchio, G., Raiteri, C.M. and Renzini, A., 1988, *ApJ* **334**, L45.
81. García López, R.J. and Spruit, H.C., 1991, *ApJ* **377**, 268.
82. Geisler, D. and Sarajedini, A., 1999, *AJ* **117**, 328.
83. Gies, D.R. and Lambert, D.L., 1992, *ApJ* **387**, 673.
84. Gilroy, K.K. and Brown, J.A., 1991, *ApJ* **371**, 578.
85. Gratton, L., 1945, *Mem. Soc. Astron. Ital.* **17**, 5.
86. Gratton, R.G., Bonifacio, P., Bragaglia, A. et al., 2001, *A&A* **369**, 87.
87. Gratton, R.G., Sneden, C., Carretta, E. and Bragaglia, A., 2000, *A&A* **354**, 169.
88. Haft, M., Raffelt, G. and Weiss, A., 1994, *ApJ* **425**, 222.
89. Hanni, L., 1984, *SvA Lett.* **10**, 51.
90. Herrero, A., 1994, *Space Sci. Rev.* **66**, 137.
91. Herrero, A., Kudritzki, R.-P., Vilchez, J.M., Kunze, D., Butler, K. and Haser, S., 1992, *A&A* **261**, 209.
92. Hill, V. and Pasquini, L., 1999, *A&A* **348**, L21.
93. Holmes, J.A., Woosley, S.E., Fowler, W.A. and Zimmerman, B.A., 1978, *AD&NDT* **18**, 305.
94. Iben, I., Jr. and Livio, M., 1991, *PASP* **105**, 1373.
95. Iglesias, C.A. and Rogers, F.J., 1996, *ApJ* **464**, 943.
96. Ivans, I.I., Sneden, C., Kraft, R.P., Suntzeff, N.B., Smith, V.V., Langer, G.E. and Fulbright, J.P., 1999, *AJ* **118**, 1273.
97. Jasniewicz, G., Parthasarathy, M., de Laverny, P. and Theévenin, F., 1999, *A&A* **342**, 831.
98. Jehin, E., Magain, P., Neuforge, C., Noels, A., Parmentier, G. and Thoul, A.A., 1999, *A&A* **341**, 241.
99. Jura, M., 1986, *ApJ* **301**, 624.
100. King, J.R., Stephens, A., Boesgaard, A.M. and Deliyannis, C.P., 1998, *AJ* **115**, 666.
101. Kippenhahn, R., in: R.J.Tayler and J.E.Hesser (eds.), *Late Stages of Stellar Evolution*, Proc. of IAU Symp. 1974, 66, Dordrecht: Reidel, p. 20.
102. Kippenhahn, R. and Weigert A., 1994, *Stellar Structure and Evolution*. Springer-Verlag, Berlin–Heidelberg-New York.
103. Kraft, R.P., 1994, *PASP* **106**, 553.
104. Kraft, R.P., Peterson, R.C., Guhathakurta, P., Sneden, C., Fulbright, J.P. and Langer, G.E., 1999, *ApJ* **518**, L53.
105. Kraft, R.P., Sneden, C., Langer, G.E. and Prosser, C.F., 1992, *AJ* **104**, 645.
106. Kraft, R.P., Sneden, C., Langer, G.E. and Shetrone, M.D., 1993,

AJ **106**, 1490.
107. Kraft, R.P., Sneden C., Smith G.H., Shetrone M.D., Langer G.E. and Pilachowski, C.A., 1997, *AJ* **113**, 279.
108. Kudritzki, R.P., Pauldrach, A., Puls, J., and Voels, S.R., 1991, in: Haynes, R. and Milne, D. (eds.), *The Magellanic Clouds*, IAU Symp. 148, Kluwer Acad. Publ., Dordrecht, pp. 279–284.
109. Kudryashov, A.D. and Tutukov, A.V., 1988, *Astron. Tsirk.* **1525**, 11.
110. Kumar, P., Narayan, R. and Loeb, A., 1995, *ApJ* **453**, 480.
111. Laird, J.B. and Sneden, C., 1996, in: Morrison, H. and Sarajedini, A. (eds.), *Formation of the Galactic Halo... Inside and Out*, ASP Conference Series **92**, p. 192.
112. Landré, V., Prantzos, N., Aguer, P., Bogaert, G., Lefebvre, A. and Thibaud, J.P., 1990, *A&A* **240**, 85.
113. Langer, N., 1992, *A&A* **265**, L17.
114. Langer, G.E. and Hoffman, R.D., 1995, *PASP* **107**, 1177.
115. Langer, G.E., Hoffman, R. and Sneden, C., 1993, *PASP* **105**, 301.
116. Langer, G.E., Hoffman, R.D. and Zaidins, C.S., 1997, *PASP* **109**, 244.
117. Langer, G.E., Kraft, R.P., Carbon, D.F. and Friel, E., 1986, *PASP* **98**, 473.
118. Langer, G.E., Suntzeff, N.B. and Kraft, R.P., 1992, *PASP* **104**, 523.
119. Lattanzio, J.C. and Frost, C.A., 1997, in: Bedding T.R. *et al.* (eds.), *Fundamental Stellar Properties: The Interaction Between Observation and Theory*, Proc. IAU Symp. 189.
120. Lattanzio, J.C., Forestini, M. and Charbonnel, C., 2000, *Mem. Soc. Astron. Italiana*, in press.
121. Lattanzio, J.C., Frost, C.A., Cannon, R.C. and Wood, P.R., in: R.F. Wing (ed.), *The Carbon Star Phenomenon*, Proc. of IAU Symp. vol. 177, Kluwer, Dordrecht, 2000, p. 449.
122. Leushin, V.V., Urpin, V.A. and Yakovlev, D.G., 1989, *Pis'ma Astron. Zh.* **15**, 1008-1015; *Sov. Astron. Lett.* **15**, 439.
123. Lyubimkov, L.S., 1984, *Astrofizika* **20**, 475; 1984, *Astrophysics* **20**, 57B.
124. Lyubimkov, L.S., 1989, *Astrofizika* **30**, 99; 1989, *Astrophysics* **30**, 58L.
125. Lyubimkov, L.S., 1991, in: Michaud, G. and Tutukov, A. (eds.), *Evolution of Stars: The Photospheric Abundance Connection*, Dordrecht, Kluwer, p. 125-135.
126. Lyubimkov, L.S., 1996, *Ap&SS* **243**, 329.
127. Maeder, A., 1983, *A&A* **120**, 113.
128. Maeder, A., 1992, *A&A* **264**, 105.
129. Maeder, A., 1995, *A&A* **299**, 84.
130. Maeder, A., 1997, *A&A* **321**, 134.
131. Maeder, A. and Meynet, G., 1996, *A&A* **313**, 140.
132. Maeder, A. and Zahn, J.-P., 1998, *A&A* **334**, 1000.
133. McWilliam, A. and Lambert, D.L., 1988, *MNRAS* **230**, 573.
134. Messenger, B., 2000, PhD Thesis, Monash University.
135. Meynet, G. and Maeder, A., 1997, *A&A* **321**, 465.

136. Nakamura, F. and Umemura, M., 1999, *ApJ* **515**, 239.
137. Norris, J., Freeman, K.C., Cottrell, P.L. and Da Costa, G.S., 1981, *ApJ* **244**, 205.
138. Norris, J.E. and Cottrell, P.L., 1979, *ApJ* **229**, L69.
139. Norris, J.E. and Da Costa, G.S., 1995, *ApJ* **447**, 680.
140. Öpik, E.J., 1951, *MNRAS* **111**, 278.
141. Paczynski, B., 1970, *Acta Astron.* **20**, 47.
142. Paltoglou, G. and Norris, J.E., 1989, *ApJ* **336**, 185.
143. Parmentier, G., Jehin, E., Magain, P., Neuforge, C., Noels, A. and Thoul, A.A., 1999, *A&A* **352**, 138.
144. Pavlov, G.G. and Yakovlev, D.G., 1978, *Astron. Zh.* **55**, 1043–1056; *Sov. Astron.* **22**, 595–602.
145. Peterson, R.C., Rood, R.T. and Crocker, D.A., 1995, *ApJ* **453**, 214.
146. Pilachowski, C.A., Sneden, C., Kraft, R.P. and Langer, G.E., 1996, *AJ* **112**, 545.
147. Pinsonneault, M., 1997, *ARA&A* **35**, 557.
148. Pinsonneault, M.H., Kawaler, S.D. and Demarque, P., 1989, *ApJ* **338**, 424.
149. Prantzos, N., Coc, A. and Thibaud, J.P., 1991, *ApJ* **379**, 729.
150. Press, W.H., 1981, *ApJ* **245**, 286.
151. Raffelt, G. and Weiss, A., 1992, *A&A* **264**, 536.
152. Ratynski, W. and Käppeler, F., 1988, *Phys. Rev. C* **37**, 595.
153. Renzini, A., 1981, in: I. Iben, Jr. and A. Renzini (eds.), *Physical Processes in Red Giants*, D. Reidel Publ. Comp., Dordrecht, p. 431.
154. Renzini, A. and Voli, M., 1981, *A&A* **94**, 175.
155. Riess, A.G., Filippenko, A., Challis, P. et al., 1998, *ApJ* **118**, 1009.
156. Ringot, O., 1998, *A&A* **335**, L89.
157. Rogers, F.J. and Iglesias, C.A., 1992, *ApJS* **79**, 507.
158. Rogers, F.J., Swenson, F.J. and Iglesias, C.A., 1992, *ApJ* **456**, 902.
159. Rood, R.T., 1973, *ApJ* **184**, 815.
160. Sackmann, I.-J. and Boothroyd, A.I., 1992, *ApJ* **392**, L71.
161. Sackmann, I.-J. and Boothroyd, A.I., 1999, *ApJ* **510**, 217.
162. Salaris, M. and Weiss A., 1997, *A&A* **327**, 107.
163. Salpeter, E.E., 1955, *ApJ* **121**, 161.
164. Schatz, H., Jaag, S., Linker, G. et al., 1995, *Phys. Rev. C* **51**, 379.
165. Schatzman, E., 1993, *A&A* **279**, 431.
166. Schatzman, E. and Maeder, A., 1981, *A&A* **96**, 1.
167. Schlattl, H., Weiss, A. and Ludwig, H.-G., 1997, *A&A* **322**, 646.
168. Shetrone, M.D., 1994, *BAAS* **26**, 1513.
169. Shetrone, M.D., 1996, *AJ* **112**, 1517.
170. Shetrone, M.D., 1996, *AJ* **112**, 2639.
171. Shetrone, M.D., in: Bedding, T.R. et al. (eds.), *Fundamental Stellar Properties: The Interaction Between Observation and Theory*, Poster Proc. of IAU Symp. 189, published by School of Physics, Univ. of Sydney, 1998, p. 158.
172. Siess, L. and Livio, M., 1999, *MNRAS* **308**, 1133.
173. Smith, G.H., 1989, in: Cayrel de Strobel, G. et al. (eds.), *The Abundance Spread within Globular Clusters: Spectroscopy of*

Individual Stars, Obs. de Paris, p. 63.
174. Smith, G.H. and Kraft, R.P., 1996, *PASP* **108**, 344.
175. Smith, G.H. and Norris, J.E., 1982, *ApJ* **254**, 149.
176. Smith, G.H. and Suntzeff, N.B., 1989, *AJ* **97**, 1699.
177. Smith, G.H. and Tout, C.A., 1992, *MNRAS* **256**, 449.
178. Smith, G.H., Shetrone, M.D., Bell, R.A., Churchill, C.W. and Briley, M.M., 1996, *AJ* **112**, 1511.
179. Smith, V.V. and Lambert, D.L., 1990, *ApJ* **361**, L69.
180. Smith, V.V., Shetrone, M.D. and Keane, M.J., 1999, *ApJ* **516**, L73.
181. Smith, V.V., Suntzeff, N.B., Cunha, K., Gallino, R., Busso, M., Lambert, D.L. and Straniero, O., 2000, *AJ* **119**, 1239.
182. Sneden, C., Kraft, R.P., Prosser, C.F. and Langer, G.E., 1991, *AJ* **102**, 2001.
183. Sneden, C., Kraft, R.P., Prosser, C.F. and Langer, G.E., 1992, *AJ* **104**, 2121.
184. Sneden, C., Kraft, R.P., Shetrone, M.D., Smith, G.H., Langer, G.E. and Prosser, C.F., 1997, *AJ* **114**, 1964.
185. Sneden, C., Pilachowski, C.A. and VandenBerg, D.A., 1986, *ApJ* **311**, 826.
186. Spite, F. and Spite, M., 1982, *A&A* **115**, 357.
187. Suntzeff, N.B., 1989, in: Cayrel de Strobel, G. *et al.* (eds.), *The Abundance Spread within Globular Clusters: Spectroscopy of Individual Stars*, Obs. de Paris, p. 71.
188. Suntzeff, N.B., 1993, in: Smith, G.H. and Brodie, J.P. (eds.), *The Globular Cluster–Galaxy Connection*, ASP Conference Series, **48**, p. 167–179.
189. Suntzeff, N.B. and Smith, G.H., 1991, *ApJ* **381**, 160.
190. Sweet, P.A., 1950, *MNRAS* **110**, 548.
191. Sweigart, A.V., 1997, *ApJ* **474**, L23.
192. Sweigart, A.V. and Mengel J.G., 1979, *ApJ* **229**, 624.
193. Takahashi, K. and Yokoi, K., 1987, *AD&NDT* **36**, 375.
194. Talon, S. and Charbonnel, C., 1998, *A&A* **335**, 959.
195. Talon, S. and Zahn, J.-P., 1997, *A&A* **317**, 749.
196. Talon, S., Zahn, J.-P., Maeder, A. and Meynet, G., 1997, *A&A* **322**, 209.
197. Thorburn, J.A., 1994, *ApJ* **421**, 318.
198. Thoul, A., Jorissen, A., Goriely, S. *et al.*, 2000, in: Noels, A. *et al.* (eds.), *The Galactic Halo: From Globular Clusters To Field Stars*, Univ. Liège, Liège, p. 567.
199. Timmes, F.X., Woosley, S.E. and Weaver, T.A., 1995, *ApJ Suppl. Ser.* **98**, 617.
200. Tout, C.A. and Pringle, J.E., 1992, *MNRAS* **256**, 269.
201. Urpin, V.A., Shalybkov, D.A. and Spruit, H.C., 1996, *A&A* **306**, 455.
202. VandenBerg, D.A., 1992, *ApJ* **391**, 685.
203. VandenBerg, D.A. and Smith, G.H., 1988, *PASP* **100**, 314.
204. Vauclair, S., 1988, *A&A* **335**, 971.
205. Ventura, P., D'Antona, F., Mazzitelli, I. and Gratton, R., 2001, *ApJ* **550**, L65.

206. Vogt, H., 1925, *Astron. Nachr.* **223**, 229.
207. Wagenhuber, J. and Weiss, A., 1994, *A&A* **290**, 807.
208. Walker, T.P., Steigman, G., Kang, H.-S., Schramm, D.M. and Olive, K.A., 1991, *ApJ* **376**, 51.
209. Wallerstein, G. and Sneden, C., 1982, *ApJ* **255**, 572.
210. Weidemann, V., 1987, *A&A* **188**, 74.
211. Weiss, A., 1994, *A&A* **284**, 138.
212. Weiss, A., Denissenkov, P.A. and Charbonnel, C., 2000, *A&A* **356**, 181.
213. Weiss, A., Keady, J.J., Magee, N.H. Jr., 1990, *AD&NDT* **45**, 209.
214. Weiss, A., Wagenhuber, J. and Denissenkov, P.A., 1996, *A&A* **313**, 581.
215. Wheeler, J.C., Sneden, C. and Truran, J.W., 1989, *ARA&A* **27**, 279.
216. Whitney, J.H. O'Connell, R.W., Rood, R.T., Dorman, B., Landsman, W.B., Cheng, K.-P., Bohlin, R.C., Hintzen, P.M.N., Roberts, M.S., Smith, A.M., Smith, E.P. and Stecher, T.P., 1994, *AJ* **108**, 1350.
217. Woosley, S.E. and Weaver, T.A., 1995, *ApJ Suppl. Ser.* **101**, 181.
218. Woosley, S.E., Fowler, W.A., Holmes, J.A. and Zimmerman, B.A., 1978, *AD&NDT* **22**, 371.
219. Zahn, J.-P., 1992, *A&A* **265**, 115.
220. Zahn, J.-P., in: Provost, J. and Schmider, F.-X. (eds.), *Sounding Solar and Stellar Interiors*, Proc. of 181st Symp. of the IAU, held in Nice, France, September 30 – October 3, 1996, published by Kluwer, Dordrecht, 1998, p. 175.
221. Zahn, J.-P., Talon, S. and Matias, J., 1997, *A&A* **322**, 320.
222. Zaidins, C.S. and Langer, G.E., 1997, *PASP* **109**, 252.
223. Briley, M., Cohen, J.G. and Stetson, P.B., 2002, *ApJ* **579**, L17.
224. Cohen, J.G., Briley, M.M. and Stetson, P.B., 2002, *AJ* **123**, 2525.
225. D'Antona, F., Caloi, V., Montalbán, J., Ventura, P. and Gratton, B., 2002, *A&A* **395**, 69.
226. Denissenkov, P.A. and VandenBerg, D.A., 2003, *ApJ* **593**, 509.
227. Denissenkov, P.A. and Herwig, F., 2003, *ApJ* **590**, L99.
228. Grundahl, F., Briley, M., Nissen, P.E. and Feltzing, S., 2002, *A&A* **385**, L14.
229. Harbeck, D., Smith, G.H. and Grebel, E.K., 2003, *AJ* **125**, 197.
230. Shetrone, M.D., 2003, *ApJ* **585**, L45.
231. Smith, G.H. and Martell, S.L., 2003, *PASP* **115**, 1211.
232. Thoul, A., Jorissen, A., Goriely, S., Jehin, E., Magain, P., Noels, A. and Parmentier, G., 2002, *A&A* **383**, 491.
233. Ventura, P., D'Antona, F. and Mazzitelli, I., 2002, *A&A* **393**, 215.
234. Yong, D., Grundahl, F., Lambert, D.L., Nissen, P.E. and Shetrone, M.D., 2003, *A&A* **402**, 985.

INDEX

alpha-process 8, 14, 16, 94, 98, 105, 119, 133, 136
aluminium problem 51, 53
angular momentum
— redistribution 13, 59, 65, 79, 144–145, 147, 157
— transport equations 64, 148
AGB stars 4, 5, 9, 16–18, 20–21, 24, 27, 40, 47–48, 52, 89, 94, 100–101, 119, 123, 125–128, 132–136, 138, 141–142, 162–164
basis models of red giants 29
Cameron-Fowler mechanism 89, 91–92, 94
CNO-cycle 4–5, 20, 29, 145, 158, 161
C-O core 5–6, 12, 123
combined scenario 14–15, 18, 20–21, 91, 95, 132, 134, 136, 139–140, 142, 162, 165
convective overshooting 6, 29, 113
core helium flash 6, 55, 71, 81, 96, 99, 105, 110, 116
differential rotation 13–14, 22, 59, 65, 79, 144, 148
diffusion model 11, 14, 19, 22, 28, 54–55, 57–58, 78, 80, 87, 161
extra-mixing 4, 6–9, 11–15, 17–23, 25–30, 32, 34–36, 40, 42, 45, 48–59, 64, 69–72, 75, 77–80, 82–90, 92–95, 97, 100, 102–110, 112, 114–116, 127, 131–134, 137–138, 140–145, 151, 160–162, 164–165
— depth and rate of 92, 115, 161
first dredge-up 5–7, 9, 25, 29, 31, 50, 56, 83–88, 96–97, 105, 111
galactic chemical evolution 6, 119
global anticorrelation 10–12, 15, 19, 23–24, 27, 40, 42–43, 45, 47, 50, 53–54, 57–58, 75, 78, 80, 85, 87, 92–93, 99, 103, 105, 106, 108, 131, 162

globular clusters 3
— star-to-star abundance variations in 56, 99, 102
— model for the chemical evolution of 21, 121, 125, 133–134, 162–163, 165
helium core 5, 11, 22, 30–31, 51, 65, 72–73, 75, 83, 95–96
hot bottom burning 5, 94, 100, 134
hydrogen burning shell 5–6, 9–13, 17, 21, 23, 25–31, 35–36, 50–53, 56, 72–74, 78, 80, 83–85, 93, 96–97, 100, 102–105, 108, 111–112, 114, 124, 128–129, 131, 133, 135, 138, 141, 161
mass discrepancies 7
meridional circulation 9, 12–13, 19, 21, 26, 29, 35, 55, 59–60, 63–64, 74–75, 77–82, 92, 102, 104, 112, 144–146, 148, 153–161, 164
metallicity 3, 7, 9, 16, 21, 25, 38, 49–50, 54, 56–57, 83, 96, 101, 105, 108, 110, 116, 120, 124, 133, 135, 141–142, 164
MgAl-cycle 20–21, 53, 161
mixing scenario 9–10, 14, 20, 26, 50, 102–103, 108, 111 114, 162
NeNa-cycle 20, 102, 109–111, 114, 132
nuclear kinetics equations 32, 56, 100–101
OB stars 7, 13, 55, 81, 143–144, 152, 164
pp-chain 5, 29, 84, 96, 100
primordial nucleosynthesis
— scenario 9–10, 14, 17–18, 20, 47, 118, 127, 132, 135–136, 141
— sites of 16–17
radiative zone 5–7, 12–13, 21, 27–30, 34–35, 50, 56, 59–60, 65–66, 70–75, 79, 84, 88, 93, 95, 100, 103–104, 138, 144–145, 147, 157

red giants
- low-metallicity 7, 9
- lithium rich 14, 19–20, 87–88

semi-convection 6, 29, 70, 145, 151

shear instability 59, 65–68, 70–71, 75, 131, 144–145

shellular rotation 13, 59, 79

s-process 20, 47, 96, 98–99, 101, 119, 127–128, 134, 164

turbulent
- diffusion 12–13, 19, 21, 55, 59, 62, 64, 67, 69–70, 74–75, 77, 79, 81–82, 92, 102, 104, 144–145, 148, 151–152, 154, 157–161, 164
- viscosity 13, 59, 64, 70, 144–145, 148, 150

type-II supernovae 119–123, 133, 136, 162–163

upper red giant branch 29, 84–85, 90–92

Zahn mechanism 14, 19, 21–22, 55, 64–65, 68, 75, 77–81, 92, 161–162, 165

www.ingramcontent.com/pod-product-compliance
Ingram Content Group UK Ltd.
Pitfield, Milton Keynes, MK11 3LW, UK
UKHW041418180426
11947UKWH00007B/189